RELATING
RANGES AND AIRSPACE
TO AIR COMBAT COMMAND
MISSIONS AND TRAINING

Albert A. Robbert   Manₗ   ...   Robert Kerchner
Willard Naslund   William A. Williams

Prepared for the United States Air Force

Project AIR FORCE

RAND

The research reported here was sponsored by the United States Air Force under Contract F49642-96-C-0001. Further information may be obtained from the Strategic Planning Division, Directorate of Plans, Hq USAF.

Library of Congress Cataloging-in-Publication Data

Relating ranges and airspace to Air Combat Command missions and training
/ Albert A. Robbert ... [et al.].
  p. cm.
  "MR-1286-AF."
  Includes bibliographical references.
  ISBN 0-8330-2934-7
  1. United States. Air Force. Air Combat Command. 2. Air bases—United States.
  3. Military reservations—United States. 4. Airspace (Law)—United States. I.
  Robbert, Albert A., 1944–

  UG633 .R393 2001
  358.4'17'0973—dc21

                                                          00-067354

RAND is a nonprofit institution that helps improve policy and decisionmaking through research and analysis. RAND® is a registered trademark. RAND's publications do not necessarily reflect the opinions or policies of its research sponsors.

The cover was prepared by Tanya Maiboroda using an image supplied by Kent Bingham, Photo/Graphic Imaging Center, Hill Air Force Base, Utah.

Published 2001 by RAND
1700 Main Street, P.O. Box 2138, Santa Monica, CA 90407-2138
1200 South Hayes Street, Arlington, VA 22202-5050
RAND URL: http://www.rand.org/
To order RAND documents or to obtain additional information, contact Distribution Services: Telephone: (310) 451-7002; Fax: (310) 451-6915; Internet: order@rand.org

Officials responsible for range and airspace management at Headquarters Air Combat Command (ACC) asked RAND's Project AIR FORCE to undertake a study that would improve the collection, evaluation, analysis, and presentation of the information needed to link training requirements to their associated airspace and range infrastructure requirements and to evaluate the existing infrastructure. This study was conducted initially in Project AIR FORCE's Resource Management Program. The work shifted to the Manpower, Personnel, and Training Program when it was formed in 1999.

This report provides findings regarding the adequacy of ACC's range and airspace infrastructure as revealed through use of an analytic structure and database assembled by RAND. A companion volume (*A Decision Support System for Evaluating Ranges and Airspace*, MR-1286/1-AF) provides information on construction, use, and maintenance of the database.

PROJECT AIR FORCE

Project AIR FORCE, a division of RAND, is the Air Force federally funded research and development center (FFRDC) for studies and analyses. It provides the Air Force with independent analyses of policy alternatives affecting the development, employment, combat readiness, and support of current and future aerospace forces. Research is performed in four programs: Aerospace Force Development; Manpower, Personnel, and Training; Resource Management; and Strategy and Doctrine.

CONTENTS

FIGURES

Training aircrews for combat requires access to ranges suitable for actual or simulated weapon delivery and to dedicated airspace suitable for air-to-air and air-to-ground tactics. To enhance this access, Air Combat Command (ACC) needs a comprehensive, objective statement of its range and airspace requirements, linked to national interests, and a means to compare existing infrastructure with these requirements.

Project AIR FORCE (PAF) and ACC, working in concert, met this need by developing an analytic structure containing the following elements:

- Operational requirements that aircrews and other combatants must be trained to support.

- Training tasks required to prepare aircrews for their assigned operational tasks.

- Range and airspace characteristics needed for effective support of each training task.

- Minimum durations of training events on ranges or airspace with specified characteristics.

- Dimensions, location, equipment, operating hours, and other characteristics of current ranges and airspace.

- Relational links among operational requirements, training requirements, infrastructure requirements, and available assets.

Elements of the analytic structure are depicted in Figure S.1. Operational missions, objectives, and tasks are referred to collectively as a *joint mission framework* (JMF)—a construct developed by PAF to be used in this study in lieu of less tractable alternatives such as commanders-in-chief's (CINCs') operational plans, units' designed operational capability (DOC) statements, the Uniform Joint Task List (UJTL), the Joint Mission-Essential Task List (JMETL), or the Air Force Task List (AFTL). Training requirements are derived primarily from ACC's Ready Aircrew Program (RAP), plus refinements identified by PAF during the course of its research. Infrastructure requirements must be expressed geographically, qualitatively, and quantitatively, along with corresponding information on existing ranges and airspace. The structure also requires a capability to match existing ranges and airspace with requirements.

RAND *MR1286AF-S.1*

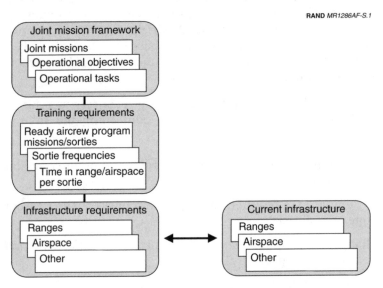

Figure S.1—The Analytic Structure

To make the information used in this analysis continuously maintainable and accessible, PAF constructed a relational database that can be used to support a variety of staff processes. It can be used, for example, to evaluate the characteristics of currently available infrastructure, determine how many sorties (by sortie type, base, and/or mission design series [MDS]) are affected by a given deficiency, determine whether current ranges and airspace provide sufficient capacity, and evaluate alternative resourcing and investment options, basing options, or training options.

During the course of our analysis, we made a number of broad observations:

- In attempting to link training requirements to national interests, we found that existing statements of operational requirements do not lend themselves to a strategies-to-task linkage to training requirements. CINCs' war plans and unit DOCs are too detailed, too context-specific, and classified at a level impractical for open communication with the public. The UJTL and its derivatives, the JMETL and AFTL, suffer from a land-centric orientation and a failure to recognize the contributions of aerospace power at strategic and operational levels of war. Consequently, we developed and linked training requirements to our own statement of operational requirements—a *joint mission framework*. The framework focuses on *effects to be achieved* for a joint commander without regard to how those needs might be met.

- Aircrew training requirements are, in many respects, not formally specified in sufficient detail to derive requirements for range and airspace infrastructure or other training resources. Most notably, we found insufficient specifications for duration of training events, the nature of simulated threats to be included in training scenarios, and requirements for training involving multiple MDSs.

- Prior to this study, centralized repositories of information on current ranges and airspace were very limited. PAF and ACC established such a system to collect relevant information for this study. The system can be expanded to record other management information regarding ranges and airspace (e.g., range and airspace utilization data), requirements for other training-related resources (e.g., flying hours, munitions, maintenance effort), and for non-ACC users (e.g., reserve components, Air Education and Training Command, Air Force Materiel Command, Headquarters USAF, or the other services).

- We found no problems in current infrastructure regarding proximity of ranges and airspace to home bases for air-to-air sorties, but there are some proximity problems for air-to-ground sorties. Insufficient size is a problem for a large proportion of military operation areas (MOAs), military training routes (MTRs), and weapon safety footprint areas (WSFAs). Deficiencies are widely observed in scoring and other feedback systems, targets, threat emitters, authorization to use chaff and flares, and terrain variety. Capacity is generally not a problem.

- To realize the power and potential of the range and airspace database, a continuing investment must be made to develop and employ the human capital needed to maintain and operate it. An appropriately trained database administrator must be assigned. Staff and field users must appreciate the system's capabilities and routinely use them.

- The decision support system has the potential to serve a much larger staff client base than was originally conceived. It can support flying training requirement analysis, base/unit beddown evaluation, and program planning.

ACKNOWLEDGMENTS

The direction and shape of this study were strongly influenced by Col Chuck Gagnon, Chief of Range, Airspace, and Airfield Management, Headquarters Air Combat Command, and Maj Michael "Buzz" Russett, our point of contact in that division, at the inception of the study. Subsequent chiefs, Cols Ron Oholendt and Lynn Wheeless, provided continuing guidance and support, as did Maj Gen David MacGhee, who was Air Combat Command's Deputy Chief of Staff for Operations at a critical point in the study. Colonel Charles Hale, Lt Col Art Jean, Lt Col Frank DiGiovanni, Lt Col Dale Garrett, Maj Rob Bray, Raul Bennett, Kent Apple, and Bob Kelchner, all within the Range, Airspace, and Airfield Management Division, also made significant contributions to the project. Kent Bingham, of the Photo/Graphic Imaging Center at Hill Air Force Base, provided a graphic image of airspace for use in our cover art. During the course of our research, we visited virtually every ACC operational wing, where local range and airspace managers, operational squadron commanders, and aircrews provided much of their valuable time arranging for our visits and participating in our interviews.

We are indebted to our Project AIR FORCE program directors, C. Robert Roll and Craig Moore, for their confidence and support. Leslie O'Neill, in RAND's Langley Air Force Base office, supported us generously and capably during our many visits to Headquarters Air Combat Command. Colleagues John Schank and John Stillion provided insightful reviews, and Jeanne Heller carefully edited the manuscript.

Responsibility for any remaining errors remains, of course, our own.

ACC	Air Combat Command
ACM	Air combat maneuver
ACMI	air combat maneuvering instrumentation
AEF	air expeditionary force
AFB	Air Force Base
AFI	Air Force Instruction
AFTL	Air Force Task List
AGL	above ground level
AHC	advanced handling characteristics
ARTCC	air route traffic control center
ATCAA	air traffic control assigned airspace
AWACS	airborne warning and control system
BFM	basic fighter maneuver
BMC	basic mission capable
BSA	basic surface attack
CAS	close air support

C^{2}ISR	command, control, intelligence, surveillance, and reconnaissance
CINC	commander-in-chief
CMR	combat mission ready
CSOT	communications system operator training
CSS	combat skills sorties
CWDS	Combat Weapons Delivery Software
DCA	defensive counter air
DOC	designed operational capability
DOR	Director of Operations, Airspace and Ranges (ACC)
FSU	former Soviet Union
G	an acceleration equal to the force of gravity (approximately 32 feet/second^2)
IMC	instrument meteorological conditions
INS	instrument
JDAM	Joint Direct Attack Munition
JMETL	Joint Mission-Essential Task List
JMF	Joint Mission Framework
JROC	Joint Requirements Oversight Council
JSTARS	Joint Surveillance and Target Attack Radar System
LFE	large force engagement
MDS	mission design series
MFCD	maximum free cruising distance
MOA	military operations area
MSL	mean sea level

MTR	military training route
NIMA	National Imagery and Mapping Agency
nm	nautical mile
OCA	offensive counter air
PAF	Project AIR FORCE
PAI	primary aircaft inventory
PFPS	Portable Flight Planning Software
PMAI	primary mission authorized inventory
RAP	Ready Aircrew Program
RPI	rated position identifier
SAT	surface attack tactics
SCDL	surveillance control data link
SEAD	suppression of enemy air defenses
SMME	small multi-MDS engagement
STP	Standard Training Plans
SUA	special use airspace
USAFE	U.S. Air Forces in Europe
UJTL	Uniform Joint Task List
UTTR	Utah Test and Training Range
WSFA	weapon safety footprint area

Chapter One

INTRODUCTION

BACKGROUND

Training aircrews for combat requires access to ranges suitable for actual or simulated weapons delivery and to dedicated airspace suitable for air-to-air and air-to-ground tactics. Air Combat Command (ACC) and other military commands responsible for training combat aircrews have access to an extensive inventory of ranges and airspace.

Faced with increasing competition for infrastructure usage, ACC recognized that it needed a requirements-based rather than a deficiency-based approach for determining its range and airspace infrastructure needs. In the deficiency-based approach that prevailed at the time, range and airspace resourcing alternatives were based primarily on statements of apparent *gaps* between requirements and existing capabilities. Better resourcing decisions could be made if both the requirements and current asset capabilities were stated more explicitly, with resourcing decisions based on rigorously derived assessments of the gaps.

To be defensible, infrastructure requirements must be linked firmly to training requirements, which in turn must be linked to operational requirements that demonstrably serve national interests. Additionally, for a requirements-based approach to succeed, an efficient means of comparing existing infrastructure capabilities with these vetted requirements is needed. RAND's Project AIR FORCE (PAF) was asked to help in developing these linked sets of requirements and assets.

The scope of the project was broad, encompassing all mission design series (MDS) aircraft operated by ACC and all of its operational unit locations. Among fighters, these include F-16s at Cannon, Hill, Moody, Shaw, and Mt. Home Air Force Bases (AFBs), F-15Cs at Eglin, Langley, and Mt. Home AFBs, F-15Es at Mt. Home and Seymour-Johnson AFBs, A-10s and OA-10s at Davis-Monthan, Moody, and Pope AFBs, and F-117s at Holloman AFB. Bombers include B-1Bs at Dyess, Ellsworth, and Mt. Home AFBs, B-2s at Whiteman AFB, and B-52Hs at Minot and Barksdale AFBs. Additionally, ACC aircrews operate a wide variety of rescue, reconnaissance, command and control, and special-mission aircraft at numerous locations.

OBJECTIVES AND APPROACH

PAF and ACC, working in concert, determined that ACC's needs could best be met through the following steps:

- Cataloging aircrew training requirements
- Relating the training requirements to operational requirements and higher-level national objectives
- Relating the training requirements to supporting range and airspace infrastructure requirements
- Comparing existing range and airspace infrastructure with requirements.

This framework called for both a repository of information on various elements and a means of representing relationships among the elements. A *relational database* was the tool of choice for meeting these needs. In addition to serving the analytic needs of this project, the database could be updated to reflect changes in requirements or existing assets or expanded as necessary to capture other related management information. In the hands of range and airspace managers at ACC or elsewhere, it could become a valuable tool for ongoing evaluation and management of range and airspace assets.

Ideally, requirements captured in the database would be developed through analysis of empirical data. Given suitable data, the impacts on operational effectiveness of training intensity and other training

environment characteristics would be demonstrated. When married with appropriate benchmarks regarding operational effectiveness, such an analytic approach would provide a robust, objective basis for training and infrastructure requirements. Unfortunately, neither detailed training data nor metrics regarding operational effectiveness, at the level of detail necessary for such analyses, are systematically or comprehensively captured. The limited data that are available are not sufficiently representative to allow generalizing across the full spectrum of training requirements. Moreover, the process of establishing the necessary metrics, capturing the training and performance data, and continually analyzing the data would be extremely, perhaps prohibitively, expensive. The alternative to such a data-driven regimen is to rely on expert judgment.

In populating the database using expert judgment, PAF sought to use the most reliable available sources and techniques. In some cases, PAF found the necessary expertise within its research staff. In other cases, the research staff relied extensively on judgments and inputs from experienced aircrews and range/airspace managers in headquarters, training, and operational units. To the extent possible, PAF enhanced the objectivity and replicability of these judgments through careful analysis of the underlying operational and training processes, including graphical representation of key tactical maneuvers. The goal was to make these judgments as visible and credible as possible inside and outside the Air Force.

ORGANIZATION OF THE REPORT

Chapter Two describes the elements of the analytic structure we adopted in our research. It also documents how PAF developed the elements and captured the information needed to populate them. In Chapter Three, we use information captured in the range and airspace database to assess the range and airspace assets used by ACC units. Chapter Four describes the capabilities of the database for ongoing analysis and outlines how it may be used to respond to the kinds of staff issues likely to be faced by range and airspace managers. The final chapter provides PAF's observations and conclusions regarding the adequacy of ACC ranges and airspace and the utility of the range and airspace database in managing them.

Chapter Two

ELEMENTS OF THE ANALYTIC STRUCTURE

We required the following elements to fully document range and airspace infrastructure requirements, trace their relevance to training and operational requirements, and assess their adequacy:

- Operational requirements that aircrews and other combatants must be trained to support.

- Training tasks required to prepare aircrews for their assigned operational tasks.

- Range and airspace characteristics needed for effective support of each training task.

- Minimum durations of training events on ranges or airspace with specified characteristics.

- Dimensions, location, equipment, operating hours, and other characteristics of current ranges and airspace.

- Relational links among operational requirements, training requirements, infrastructure requirements, and available assets.

These elements relate to each other in an analytic structure that is depicted in Figure 2.1. Operational missions, objectives, and tasks are referred to collectively as a *joint mission framework*. As the figure implies, infrastructure requirements, training requirements, and the joint mission framework must be serially linked. Additionally, infrastructure requirements and current infrastructure must be linked in a way that permits ready comparisons.

RAND *MR1286AF-2.1*

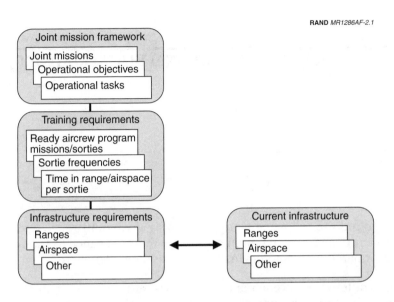

Figure 2.1—The Analytic Structure

We next describe the elements of the range and airspace analytic structure and document how we developed and populated the various elements and linkages.

OPERATIONAL REQUIREMENTS: THE JOINT MISSION FRAMEWORK

This tier in the analytic system identifies operational requirements essential to national defense and demonstrates how each aircrew training requirement is linked to one or more of the operational requirements. In developing this tier, we first examined existing representations of national defense needs: war plans developed by joint commanders in chief (CINCs), unit designed operational capability (DOC) statements, the Uniform Joint Task List (UJTL), Joint Mission-Essential Task Lists (JMETLs), and the Air Force Task List (AFTL). For reasons discussed below, we did not find an acceptable structure among these available frameworks. Instead, we built upon earlier RAND work devoted to identifying joint missions (Kent and

Ochmanek, 1998; Thaler, 1993; Pirnie and Gardiner, 1996) and developed our own *joint mission framework*.

Existing Representations of National Defense Needs

War Plans and DOCs. War plans and DOCs have several characteristics that militated against their use to represent operational requirements in our analytic structure. War plans are unique to various theaters and DOCs are unique to various units. Collecting and organizing all war plans and DOCs within a common reference system would present a massive task. Additionally, the relationship of DOCs to war plans is not explicit; it may not be possible to identify specific linkages. Finally, the resulting product would require a security classification that would preclude its use for expressing range and airspace needs to the public or within some areas of the training community.

The UJTL. Derived from joint doctrine, the UJTL is a joint framework that provides one basis for how services provide their capabilities and how training can be shaped to support these capabilities. It is written at three levels of military endeavor (strategic, operational, and tactical). It is a tool for the unified CINCs to declare their mission priorities to the national command authority. As an input to the Joint Requirements Oversight Council (JROC), the UJTL is a significant shaper of resources.

The UJTL process has some shortcomings, however. UJTL tasks are not statements of effects to be achieved by forces but rather tend to specify the means to be used to obtain the effects. The UJTL is redundant as it progresses through its three levels. Further, it has a ground conflict model as its principal basis, one that leaves out the significant potential of parallel warfare and the capabilities of the services—mostly those of the Air Force—to affect the entire spectrum of conflict and nonconflict. Finally, it allows the specific contributions of the services to enter the framework only at the tactical task level, where capabilities are depicted primarily as supporting ground combat objectives rather than contributing to the national military objectives of the overall campaign.

JMETLs. The JMETLs are priority listings of tasks required by various CINCs to execute their war plans. Unfortunately, they do not provide

a comprehensive view of what a service could do to support a CINC except at the highest priority levels.

The AFTL. The AFTL is the Air Force's elaboration of the UJTL. It is constrained by the UJTL framework and its focus on the ground campaign. It does not provide active, operational statements of how air and space power can support the full spectrum of national objectives. The statements are directive with respect to means and in extreme detail, with little room for the creative application of military force. Again, like the UJTL, they presuppose means instead of focusing on a statement of needed effects.

The Joint Mission Framework

In developing this framework, we sought to express the CINCs' needs in terms of desired operational effects rather than in terms of the processes used to achieve them. The Air Force for many years has defined its capabilities as classic "missions," such as close air support or interdiction. More recently, it has defined its contributions to warfighting as a series of core capabilities: air and space superiority, precision engagement, rapid global mobility, global attack, information superiority, and agile combat support. These mission or capability statements generally describe processes—means to various operational ends. For our framework, we sought to describe the operational ends themselves. We also wanted a framework that would allow us to relate airpower capabilities directly to strategic and operational as well as CINC-derived tactical objectives.[1]

Using a strategies-to-tasks concept, we developed a set of operational missions, objectives, and tasks to describe how military power can be applied jointly. The framework, found in Appendix A, contains 11 joint operational missions that collectively describe the broad outcomes CINCs seek to achieve in operations ranging from major theater war to smaller-scale peacekeeping and peacemaking contingencies. Within these missions, we identify some 40 operational objectives and 150 operational tasks.

---

[1]The Gulf War (Operations Desert Shield and Desert Storm) and the more recent air war over Serbia (Operation Allied Force) demonstrated that air and space capabilities are powerful instruments that can be used independently of a ground campaign to achieve many operational and strategic objectives.

An important feature of the framework is its simplicity and clarity. The statements of desired effects and the worth of achieving the effects should be easily understandable to a wide range of Air Force and non–Air Force audiences.

TRAINING REQUIREMENTS: AN ADAPTATION OF THE READY AIRCREW PROGRAM

The next element in the analytic structure represents training activities needed to prepare aircrews to support operational requirements. To complete the linkages envisioned in the analytic structure, training activities must be related, on one hand, to operational requirements, and on the other hand, to training resource needs, specifically range and airspace infrastructure.[2]

Flying training may be divided into several categories (Hq. U.S. Air Force, AFI 11-202, Vol. 1, pp. 6, 9, 13, 16):

1. *Undergraduate flying training* to provide basic flying proficiency

2. *Initial qualification training* to qualify for basic aircrew duties in an assigned position for a specific MDS aircraft

3. *Mission qualification training* to qualify in an assigned aircrew position to perform a command or unit mission

4. *Continuation training* to provide the volume, frequency, and mix of training necessary to maintain proficiency at the assigned qualification level

5. *Special mission training* to provide any special skills necessary to carry out the unit's assigned missions that are not required by every crew member

6. *Upgrade training* to prepare aircrew members to perform as flight leads, instructor pilots, mission commanders, or other advanced roles.

---

[2]Readers should not infer that training requirements used in our analysis were derived from our joint mission framework. We derived our training requirements from the Air Force's Ready Aircrew Program (RAP), as described below, which in turn is derived from other representations of operational requirements such as unit DOCs. We then linked our training requirements framework to our joint mission framework.

Of these types, we focused primarily on mission qualification and continuation training. Undergraduate flying training and initial qualification training are accomplished primarily through formal training courses and generally do not place demands on ACC range and airspace infrastructure.[3] Special mission and upgrade training is often accomplished using sorties that are dual-logged as continuation training. Thus, demand for ACC ranges and airspace is largely a function of mission qualification and continuation training requirements.

Mission qualification and continuation training requirements are outlined in the Air Force's Ready Aircrew Program. RAP requirements are contained in MDS-specific, 11-2 series Air Force Instructions (AFIs) and in annual tasking messages published by ACC. For aircrews in each MDS, RAP specifies a total number of sorties per training cycle, broken down into mission types, plus specific weapons qualifications and associated events. The specified number of sorties varies depending on the aircrew member's experience and qualification level.[4] For example, mission category sorties for one type of F-16 for the 1998–1999 RAP cycle are shown in Table 2.1.

---

[3]For a few weapon systems, ACC does conduct initial qualification training. These programs usually are co-located with at least one combat squadron and must share local training infrastructure. This study did not include the initial training requirements for these systems; therefore, the total requirement for these bases is underestimated.

[4]*Experienced* pilots have accumulated a specified number of flying hours. For example, fighter pilots are considered experienced if they have accumulated 500 hours in their primary aircraft, or 1000 total hours of which 300 are in their unit's primary aircraft, or 600 fighter hours of which 200 hours are in their unit's primary aircraft, or who reached an experienced level in another fighter MDS and have 100 hours in their unit's primary aircraft.

Line pilots in operational units generally attain a qualification level designated *combat mission ready* (CMR). Pilots in staff positions generally attain a lower level of qualification designated *basic mission capable* (BMC).

In each training cycle, RAP specifies more sorties for inexperienced aircrew members than for experienced aircrew members and more sorties for CMR qualification than for BMC qualification. Additionally, RAP may specify more sorties for active component aircrews than for reserve component aircrews.

Table 2.1

Ready Aircrew Program Mission Category Sorties for the F-16CG

| Mission Category | Annual Sortie Requirement | | | |
| | Basic Mission Capable | | Combat Mission Ready | |
	Inexperienced	Experienced	Inexperienced	Experienced
Basic surface attack				
(BSA)(day)	6	4	8	6
(BSA (night)			4	3
Surface attack tactics				
(SAT) (day)	6	4	14	12
(SAT (night)			4	3
Close air support (CAS)			4	3
Defensive counter air				
(DCA) (day)	3	2	10	8
(DCA (night)			4	2
Air combat maneuver				
(ACM)			8	6
Basic fighter maneuver				
(BFM)	3	2	8	6
Red air (opposing force for air-opposed training sorties)			8	8
Commander option	54	48	18	19
Total	72	60	90	76

Sortie Types Used in the Analysis

RAP sorties may be either *basic* or *applied*. Basic sorties are building-block exercises, such as advanced handling characteristics (AHC), basic surface attack, or basic fighter maneuver, that are used to train fundamental flying and operational skills. Applied sorties, such as surface attack tactics and defensive counter air, are intended to more realistically simulate combat operations, incorporating intelligence scenarios and threat reaction events.

The examples of sortie types in the preceding paragraph are all fighter-oriented. For nonfighter aircraft, basic sorties are generally identified as combat skills sorties (CSS). Applied sorties for non-fighter aircraft are generally identified as SAT sorties (bombers) or *mission* sorties (for aircraft that do not deliver weapons).

For our analysis, we generally used the RAP sortie structure and annual sortie requirements as a statement of training requirements. However, in some cases, notably SAT, we subdivided RAP sorties into several types (which we refer to as *variants*) that differ significantly from each other in their infrastructure requirements. For example, fighter SAT missions are divided into *air opposed, ground opposed,* and *live ordnance* variants. Similarly, SAT missions for bombers are divided into *inert high/medium level, inert low level, live ordnance, simulated delivery of ordnance,* and *maritime* variants.

We also postulated the need for sorties that combine several MDS, performing different operational roles, in a single training mission. For squadron-size exercises, we used the term *large force engagement* (LFE) to identify these sorties. For less than squadron-size exercises, we used the term *small multi-MDS engagement* (SMME). We refer to LFEs and SMMEs collectively as *combined* sorties. We did not develop a comprehensive list of SMMEs. However, we have structured several examples that suggest the possibilities of specifying such training requirements. In general, RAP does not currently specify multi-MDS sorties except for LFE requirements in some MDS and a few exceptions such as FAC-A, which requires a forward air control aircraft working in conjunction with close air support aircraft.[5]

A complete list of the sortie types used in our analysis and their categories and definitions can be found in Appendix B.

Relating Training Requirements to Operational Requirements

We determined that applied and combined sorties would be related directly to operational tasks found in our joint mission framework. Basic sorties and variants would be related to various applied sorties, and could then be related indirectly to operational tasks. The

---

[5]In our base visits, we found evidence that squadrons did try to build multiple MDS sorties when they had access to other MDS aircraft and when the tactic being trained called for it. In most cases, these sorties are logged as regular applied mission sorties. To the degree that RAP does not *require* this activity, there is the danger that RAP-based calculations could underestimate actual airspace requirements.

relationships are shown in Figure 2.2. Matrices relating various MDS/sortie combinations to specific operational tasks within the joint mission framework are too large to be readily included here. However, the linkages are reflected in the range and airspace database we constructed and can be extracted for any MDS, sortie type, or joint mission. As an example, the joint mission, "Deny the enemy the ability to operate ground forces," contains an operational objective, "Halt invading armies," within which one of the operational tasks is "Delay/destroy/disrupt lead units of invading armies." For the F-16CG, the database associates this operational task with three types of applied sorties (CAS, DCA, and LFE) and five types of basic sorties (instrument [INS], AHC, ACM, BFM, and BSA). Range and airspace infrastructure required for the F-16CG for each of these sortie types can also be extracted from the database and linked to this operational task.

REQUIRED INFRASTRUCTURE CHARACTERISTICS

The next element in our analytic system is a statement of the range and airspace infrastructure needed to support training requirements. To be useful for training, the range and airspace infrastructure must have certain geographical, qualitative, and quantitative characteris-

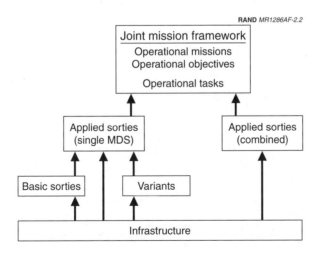

Figure 2.2—How Sorties Link Infrastructure to Operational Tasks

tics. Geographically, it must be reasonably proximate to base operating locations. For many MDS, especially fighters, extending aircraft range through air refueling is not a viable option for training sorties.[6] Even for longer-legged bomber and command, control, intelligence, surveillance, and reconnaissance (C^2ISR) aircraft, other constraints such as crew duty day length and flight time engaged in useful training versus time spent cruising to and between training areas need to be considered. Qualitatively, the infrastructure must have minimum dimensions, equipment, authorization for operating aircraft and systems in specified ways, and other characteristics. Quantitatively, the time available on proximate ranges and airspace must be sufficient to support the training requirements at an operating base. In this and the following sections, we discuss how these infrastructure requirements were developed and are represented in the range and airspace information system.

Distance from Base to Range/Airspace

Ranges and airspace must be reachable with the maximum fuel load consistent with the sortie type. Further, fuel available for cruising to, from, and between ranges and airspace must take into account the amount of fuel consumed during training events.[7] Because many sortie types require access to more than one asset (e.g., a low-level route, a maneuver area, and a range) during a given sortie, the required geographical proximity of the assets cannot be adequately expressed in terms of a radius from the base. It is better expressed in terms of a maximum for the sum of the free cruising legs between assets (see Figure 2.3). We calculated this maximum for each MDS/sortie-type combination and used it to analyze the geographical relationships of bases, ranges, and airspace.

---

[6]Air refueling should be scheduled in accordance with the need to be proficient in that skill, but air refueling assets are too limited to be used routinely to extend the training range of fighters.

[7]We use the term *training event* to indicate a part of a sortie with a specific training focus. For example, an air-to-ground sortie may include a low-level navigation leg, a threat evasion exercise, and a series of weapon deliveries. In our usage, these training-related components of the sortie are referred to as a training event.

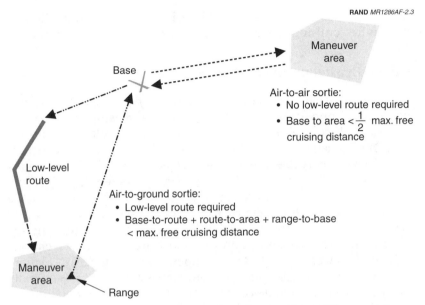

RAND *MR1286AF-2.3*

Figure 2.3—Maximum Distance from Base to Range/Airspace

To calculate the maximum free cruising distance for each MDS/sortie-type combination, we interviewed aircrew members to determine the normal external fuel-tank configuration and fuel capacity for the sortie; fuel consumption for taxi, takeoff, and climb; fuel consumption during training events en route, in the area, and/or on the range; and reserve fuel requirements. To determine fuel consumption, we first determined standard minimum durations for each training event. Minimum durations are not currently specified in ACC or Air Force training specifications. We determined reasonable values for these minimums through consultation with experienced aircrew members.

Subtracting required consumptions and reserve from fuel capacity yields the amount of fuel that can be used for free cruising legs. Dividing this amount by an average fuel consumption rate at a typical cruising speed and altitude yields the maximum free cruising time. Multiplying this time by the typical cruising speed gives the maximum free cruising distance. Maximum free cruising distances ranged from 79 miles for F-15C BFM sorties to 1757 miles for B-52

SAT sorties. In general, fighters and helicopters are far more limited in their free cruising distances than bombers and C^2ISR platforms.

Qualitative Requirements

Most qualitative infrastructure requirements (e.g., range dimensions, equipment, operating authorizations) were developed through a multistep process. First, a series of panels were conducted—one for each MDS—using ACC staff members with recent aircrew experience. The panels developed a requirements template for various mission types.[8] Second, PAF disaggregated the requirements to a training event level. Third, for those systems represented at the USAF Weapons School, instructors from the school were asked to review and revise the event-level infrastructure requirements. Fourth, the requirements information obtained from Weapons School visits were brought to each operational wing in ACC, where they were reviewed by one or more aircrew members—generally Weapons School graduates or other highly experienced personnel—in each MDS flown by the wing. Finally, we pooled the judgments gathered in this series of visits to construct an infrastructure requirement for each MDS/sortie combination.

In some cases, we noted that infrastructure requirements would vary significantly depending on the kinds of weapon deliveries or other training events included in the sortie. This fact argued for using the training event rather than the sortie type as the unit of analysis for infrastructure requirements. However, two other factors argued against using the training event as the unit of analysis. First, we did not believe that a count of required training events could be derived from RAP or any other source.[9] (A count of requirements is needed to quantify demand for ranges and airspace.) Second, the sheer number of various types of training events made this approach infeasible.

---

[8]This effort preceded the development and implementation of RAP, but the results were later harmonized with RAP sortie types.

[9]Required frequencies for some critical events are specified in RAP tasking messages or AFI 11-series publications; however, frequency requirements for most events are not specified.

In general, we used the MDS/sortie-type combination as the unit of analysis (i.e., each MDS/sortie-type combination would have its own unique set of infrastructure requirements). However, where choice of events would significantly alter the infrastructure requirement, we divided the RAP sortie into two or more variants, as discussed previously. This enabled us to better reflect specific infrastructure standards for the wide variety of crew activity logged under any one sortie type.

As an exception to the general rule of using the MDS/sortie-type combination as the unit of analysis, we found that for range characteristics related to weapon deliveries, it was necessary to use the training event (i.e., the weapon delivery type) as the unit of analysis. Weapon deliveries can vary by release altitude, release type (level, loft, dive, etc.), weapon type (rocket, gravity bomb, guided munition, etc.), level of threat (which affects assumed delivery accuracy), and MDS.[10] Weapon delivery type affects two categories of range characteristic requirements—restricted airspace dimensions and weapon safety footprint area (WSFA) dimensions. To specify standard range requirements for weapon deliveries at an MDS/sortie-type level of analysis, we would have to identify the most demanding (in terms of these range characteristics) weapon deliveries that aircrews should routinely employ in each MDS/sortie combination. However, we found no basis for selecting which weapon delivery types should be used to set these requirements. Thus, restricted airspace and weapon safety footprint area requirements are expressed at the event rather than the MDS/sortie-type level in our analysis.

Organizing the Qualitative Requirements

Qualitative requirements (and corresponding information on existing assets) were captured for six infrastructure types: low-level routes, maneuver areas, ranges, threats, orbits, and other. Specific characteristics appearing in these requirement arrays are listed in Appendix C.

---

[10]ACC currently identifies 210 distinct weapon delivery types.

This organization was developed to state the need for infrastructure without being limited to current airspace terms such as restricted area, military operation area (MOA), warning area, air traffic control assigned airspace (ATCAA), or military training route (MTR). These terms are for the most part derived from the air traffic control lexicon rather than a training lexicon. Moreover, training requirements can often be met by any of several current airspace types, or, as is frequently observed, they may require combinations of several airspace types. Thus, we sought to define the infrastructure requirements using more generic terms, e.g., low-level route rather than MTR, maneuver area rather than MOA.

Low-Level Routes. Air-to-ground sorties are generally required by training publications (AFI 11-2 series) to incorporate a low-level ingress route. An MTR typically connects to a MOA surrounding a range. The length of the route, its required altitudes, and other required attributes are captured in the range and airspace database.

Maneuver Areas. Air-to-ground sorties may require controlled airspace for attack tactics and threat reaction, generally requiring a MOA and perhaps a vertically adjacent ATCAA. Air-to-air sorties also require a maneuver area—either a MOA with an ATCAA or an offshore warning area. Required vertical and lateral dimensions and other attributes of the maneuver area are captured in the range and airspace database.

Required dimensions of these maneuver areas depend to a great extent on the aircraft maneuvers expected to be conducted within them. While we relied heavily on the expertise of experienced aircrews to specify these requirements, we developed graphical analytic tools to aid in the process. These tools are illustrated in Appendix D.

Ranges. A range is required for air-to-ground sorties. Ranges also require restricted airspace over their targets large enough to contain released weapons and the long and cross dimensions of weapon safety footprints. Required vertical and lateral dimensions of the restricted area, types of targets, scoring systems, and other related range attributes are specified in the range and airspace database. The relationship of weapon safety footprints (the long, short, and cross dimensions for a given weapon delivery), WSFAs (the area

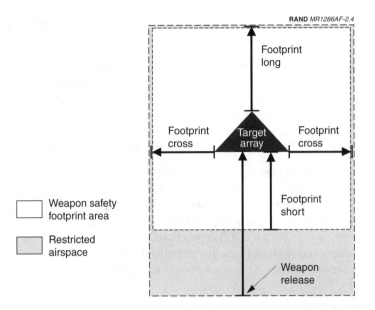

RAND *MR1286AF-2.4*

Figure 2.4—Weapon Safety Footprints, Weapon Safety Footprint Areas, and Restricted Airspace

covered by the footprint plus the size of the target array), and restricted airspace is illustrated in Figure 2.4.[11,12]

Threats. Many air-to-ground sorties require ground-based radar threat emitters or communications jammers, which may be installed on a range, beneath a MOA, or conceivably at points along an MTR. We determined that the training requirement would be met if the threat emitters were installed in any of these locations. Thus, rather than include threat requirements within range, area, and route

---

[11]To calculate WSFA and restricted airspace requirements, RAND used (1) weapon safety footprint data for 210 distinct delivery types, obtained from ACC/DOR in August 1999, (2) an assumed target array size of 2 nm × 2 nm, and (3) weapon release points calculated using Combat Weapons Delivery Software (CWDS) provided by the Mission Planning Support Facility, OO-ALC/LIRM, Hill AFB, UT.

[12]Figure 2.4 provides the WSFA and restricted airspace requirements for only a single axis of attack. For multiple axes of attack, the dimensions shown in the figure must be rotated around the target.

requirements arrays, we established a separate threat requirements array in the range and airspace database.

Orbits. Orbits may be required for air refueling or certain command and control missions. If so, the requirement is captured in the range and airspace database. Generally, we found the need for orbits as part of training events to be underdocumented. Orbits can be flown in a MOA or ATCAA, but are usually specified only in a letter of agreement with the affected air route traffic control center (ARTCC).[13]

Other. Some sorties require a specific other aircraft for effective training. For example, DCA and offensive counter-air (OCA) sorties require *red air* opponents. Others require an air or ground weapons director. Requirements such as these are not, strictly speaking, part of the range or airspace infrastructure. However, in the interest of more completely documenting training requirements, we collected such non-infrastructure requirements that came to our attention.[14]

Capacity

The amount of operating time required on ranges and in airspace can be calculated, for a given MDS/sortie-type combination, by multiplying the required number of sorties by the time required for an individual sortie on a range and/or in an airspace. After certain adjustments (discussed below), the results can be summed across all MDS/sortie-type combinations to determine a base's total local demand for ranges and airspace (referred to as assets). This demand is computed and recorded in the range and airspace database for

---

[13]Every air-to-air unit we talked to agreed that flying some sorties with an airborne warning and control system (AWACS) controller is essential, but neither the AWACS nor fighter RAP requirements list this need. Likewise, air-to-ground fighter aircrews should train occasionally with joint surveillance and target attack radar system (JSTARS) crews. As a consequence, the letters of agreement with local ARTCCs should establish orbits allowing AWACS and JSTARS aircraft to be in the right position for training with fighters in MOAs. Unfortunately, these requirements are often overlooked when fighter units negotiate new airspace agreements. Airspace managers for AWACS, JSTARS, and fighter units would benefit from closer coordination. The first step would be to establish a *requirement* that these communities train together.

[14]Pilots we interviewed said that training with other MDS is very important, but the lack of a *requirement* for such training often discouraged an already-busy potential "partner MDS" from participating in such training.

each base/MDS/sortie-type combination. In the following paragraphs, we discuss, first, how the required number of sorties is calculated and, second, how the time required for each sortie is determined.

Required Number of Sorties. The database contains a table that lists the total number of annual sortie requirements by base, MDS, and sortie type. To populate this table, we determine the number of pilots in each MDS at each base and multiply that number by the annual requirement for each sortie type.[15] The required calculations are shown in Figure 2.5 and described below.

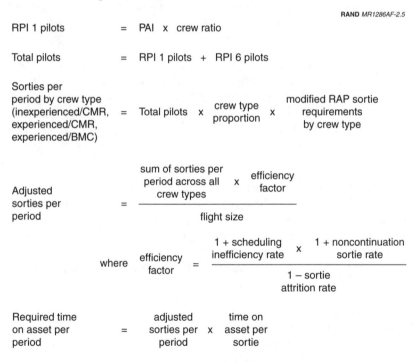

RAND *MR1286AF-2.5*

Figure 2.5—Determining Sortie and Time on Asset Requirements

---

[15]In some MDS, crew positions other than pilot also require training. However, we found no MDS with a crew position that required more sorties than the pilot. Thus, using pilot counts alone (excluding co-pilots) as the basis for annual sortie requirements is sufficient to establish an upper bound on sortie demand.

To determine the number of pilots, we first obtain the primary mission authorized inventory (PMAI) by MDS and base. These counts are multiplied by the crew ratio for the MDS, yielding the expected number of RPI 1 (RPI = rated position identifier) pilots on the base.[16] To this number, we add the number of RPI 6 pilots by base and MDS.[17] The total number of pilots is then distributed to experienced/inexperienced and BMC/CMR categories.[18]

The next step in determining the total sortie requirement is to multiply the number of pilots by the number of annual sorties required in each MDS/sortie-type combination. The number of sorties in each training cycle (generally one year) for experienced/inexperienced and BMC/CMR categories is specified by sortie type and MDS in annual RAP tasking messages.

For our analysis, we modify the raw RAP counts in several ways. We use assumed rates to redistribute RAP sortie counts to our modified-RAP variants. Additionally, we distribute commander's option sorties to specific sortie types in the same proportions that the specific sorties had relative to each other; i.e., if SAT sorties are 40 percent of the noncommander's option sorties, we distribute 40 percent of the commander's option sorties to SAT.[19]

The next step in computing the sortie requirement is to adjust for flight size. When two-ship or four-ship flights use a range or airspace, multiple aircrews obtain training in the same time period. Thus, the critical factor in quantifying range and airspace demand is not the annual number of sorties but rather the annual number of flights. To convert sortie counts to flight counts, we divide sortie

---

[16]RPI 1 identifies line pilots (excluding commander and operations officer) occupying cockpits in operational squadrons.

[17]RPI 6 identifies commanders, operations (ops) officers, and pilots in staff positions.

[18]For these calculations, we consider RPI 6 positions, except commander and ops officer, to be experienced and BMC. Commander and ops officer are considered experienced and CMR. RPI 1 pilots are considered CMR and are distributed using assumed rates between experienced and inexperienced categories.

[19]RAP specifies the number of sorties by type that each aircrew member must fly in a training cycle. Additionally, it specifies a number of sorties that can be of any type, depending on the commander's judgment of where the individual or unit needs training emphasis.

counts by an assumed average flight size for each MDS/sortie-type combination.

The final step in developing and adjusting the sortie requirement is to inflate the count to account for attrition (maintenance and weather cancellations), scheduling inefficiency, and noncontinuation training sorties. Some scheduled sorties cannot be completed because of either maintenance or weather aborts. Although these aborted sorties do not satisfy training requirements, they nonetheless consume available time on ranges and airspace because the scheduled time generally cannot be reallocated on short notice (in the case of maintenance aborts) or used by other aircrews (in the case of weather or mission conflict aborts).[20] A scheduling inefficiency factor accounts for the fact that perfectly efficient scheduling, using 100 percent of available range or airspace time, would tend to suboptimize overall aircrew time management because it would adversely affect aircrew workday and work/life balance considerations. Finally, some but not all upgrade and special qualification sorties are dual-logged as RAP sorties. The noncontinuation training inflation factor builds a range/airspace infrastructure requirement for upgrade and special qualification sorties that are not dual-logged. The range and airspace database uses assumed values for these three factors (10 percent for each factor).

Time Required per Sortie on Range and/or in Airspace. A table indicating time required per sortie on a range or in an airspace, by MDS and sortie type, is found in the database (see, for example, times indicated in Table 3.2 in Chapter Three). The times shown in Table 3.2 (minimum training event durations) are assumed values based on interviews with Weapons School and operational unit aircrews. They represent minimums considered necessary for the sortie to produce some standardized training value.[21]

---

[20]A few units fly a large number of sorties on ranges that they do not control, which can result in a mission conflict and loss of scheduled training time. Usually, once a non-owning unit arranges for time on a range, there is little chance of mission conflicts with the owning unit. However, we found at least one range (White Sands Missile Range Complex) where the range time could be canceled by range controllers within 15 minutes before entry time. In this case, fighter aircraft are already airborne when they are canceled.

[21]Two of the assessments performed in the database and described in Chapter Three are sensitive to these assumed minimum training event durations. Geographical

Total Demand. Total range and airspace time requirements by base, MDS, and sortie type are calculated and reflected in a table in the database. Table 2.2 reflects, for example, an extract of this part of the database for F-16CGs at Hill AFB. This requirement can be interpreted as a demand for maneuver airspace time for air-to-air sortie types and as a demand for both maneuver airspace and range time for air-to-ground sortie types. It is determined, as shown in Figure 2.5, as the product of total requirements for a given base/MDS/sortie-type combination multiplied by the time required on asset for that MDS/sortie-type combination.

Data Limitations. Lack of available empirical data and other related problems required us to estimate many of the factors used to compute capacity requirements. A discussion of these limitations is provided in Appendix E.

Table 2.2

Infrastructure Demand: F-16CGs at Hill AFB

Sortie Type	Total Sorties	Time per Sortie (minutes)	Average Flight Size	Required Infrastructure Time (hours)
BFM	752	40	1	674
BSA	1,128	40	2	506
CAS	357	50	2	200
DCA	1,203	35	4	236
SAT	1,474	35	4	289
SEAD-C	184	30	4	31

NOTE: SEAD = suppression of enemy air defenses.

(proximity) assessments are sensitive to them because fuel available to cruise to and from training areas is a function of fuel consumed in training events. Quantitative (capacity) assessments are sensitive to them because the total time required in a range or airspace is a sum of the times required on each sortie. It is useful to examine *how* sensitive our findings are to the assumed values we used for minimum training event duration. As will be reported in Chapter Three, we encountered few actual proximity or capacity constraints, so shortening minimum training event durations would not significantly change the results. Lengthening the minimums would increase proximity deficiencies in many cases, but would not have much effect on capacity because most installations have abundant slack capacity.

CURRENT INFRASTRUCTURE

Information regarding the characteristics of ranges and airspace commonly used by ACC aircrews was collected (by e-mail) by ACC/DOR during late 1998 and early 1999. Preformatted Excel spreadsheets were sent as attached documents to local range managers and airspace schedulers, who entered the required information in the spreadsheets and returned them to ACC/DOR. The spreadsheets were subsequently forwarded to PAF to be incorporated in the database. Subsequently, a capability was provided to permit local range managers and airspace schedulers to update these characteristics via a web interface. Specific characteristics tracked in the range and airspace database are listed in Appendix C. They can be found in various tables in the database and in selected displays available via a web browser. Limitations on the available data are discussed in Appendix E.

COMPARISON OF CURRENT INFRASTRUCTURE WITH REQUIREMENTS

An important element of our analytic structure is a capability to compare requirements and resources. Linkages and models embedded in the range and airspace database permit current infrastructure and requirements to be compared for each MDS/sortie-type combination. These comparisons are reflected in a series of tables in the database and in a display accessible via a web browser. The example from the web browser shown in Figure 2.6 depicts an assessment of maneuver areas for F-15C DCA sorties. Each row represents a different maneuver area (identified in the "name" column). Characteristics of the various areas are shown under "width," "length," etc. Characteristics that meet requirements are shaded light gray (green on the web) while those that do not meet requirements are shaded dark gray (red on the web). This screen depicts only part of a much larger matrix containing all areas and all characteristics of areas.

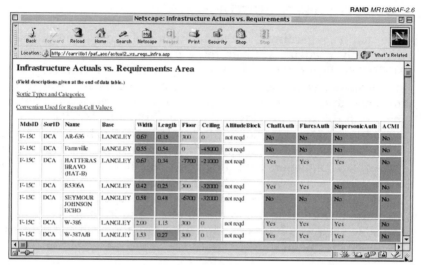

Note: Entries in width and length columns indicate the proportion of the requirement met by dimensions of the asset. Entries in floor and ceiling columns indicate the difference between requirement and dimensions of the asset. "Yes/no" entries in other columns indicate whether or not required characteristic is available on the asset.

Figure 2.6—Web Page Comparing Current Assets with Requirements

Chapter Three

ASSESSMENT OF RANGES AND AIRSPACE

In this chapter, we use the range and airspace database to provide a current assessment of the assets used by ACC units. We assess these assets in three ways:

Geographically—are the assets close enough to home bases to permit minimum required duration of training?

Qualitatively—do the assets have the standard features required for the types of sorties flown in or on them?

Quantitatively—do the operating hours of the assets provide sufficient time capacity to accommodate the required number of sorties?

DISTANCE OF ASSETS FROM HOME BASES

For ranges and airspace to be useful, they must be in proximity to the home bases of the aircraft that use them. This is especially important for fighters, which have relatively short unrefueled ranges and for which aerial refueling on routine training missions is not a practical option. Accordingly, for fighters, we calculated the maximum free cruising distances between assets (as illustrated in Figure 2.3 and described in the accompanying text). These *standard distances*, by sortie type and MDS, are shown in Table 3.1. The standard minimum training event times used in calculating standard distances are shown in Table 3.2.

Table 3.1

Maximum Free Cruising Distances (in nm) for Fighter Training Sorties

Sortie Type		A/OA-10	F-15C	F-15E	F-16CG	F-16CJ	F-16GP
General	AHC	283	219	485	172	172	172
Air-to-air	ACM	283	146	218	88	88	88
	BFM	283	79	145	88	88	88
	DCA		209	242	88	88	88
	OCA		209	242	88	88	88
	OCA_ANTI-HELO	144					
Air-to-ground	BSA	144		348	247	247	247
	SAT	144		222	100	100	100
	FAC-A	144					
	CAS	144		222	100	100	100
	SEAD					349	
	SEAD-C				100		100

NOTE: OCA = offensive counter air.

Table 3.2

Minimum Training Event Durations (in minutes) for Fighter Training Sorties

Sortie Type		A/OA-10	F-15C	F-15E	F-16CG	F-16CJ	F-16GP
General	AHC	45	25	25	30	30	30
Air-to-air	ACM	45	40	40	40	40	40
	BFM	45	40	40	40	40	40
	DCA		35	35	35	35	35
	OCA		35	35	35	35	35
	OCA_ANTI-HELO	65					
Air-to-ground	BSA	65		65	60	60	60
	SAT	65		50	50	50	50
	FAC-A	65					
	CAS	65		50	50	50	50
	SEAD					50	
	SEAD-C				50		50

NOTE: Indicated duration is sum of time required on low-level route (if any) and time required in maneuver area.

In evaluating free cruising distances for combinations of bases, maneuver areas, low-level routes, and ranges, we found that the geographical data we captured in the range and airspace database were inadequate. The database contains geographical coordinates (longitude and latitude) for the various bases, the entry points of low-level routes, and the *center* of maneuver areas. To properly evaluate the distances, we needed to know the locations of alternate entry and exit points for routes and the *edges* of maneuver areas closest to bases and route exit points. To make these evaluations, we used information from the *Digital Aeronautical Flight Information File* (DAFIF) (National Imagery and Mapping Agency [NIMA], 1999), as viewed using FalconView (computer mapping software), and using features of Portable Flight Planning Software (PFPS) (developed by the 46th Test Squadron, Eglin AFB, FL).

Maneuver Areas for Air-to-Air Sorties

For air-to-air sorties, a maneuver area is considered available within the standard distance if its distance from a base is no more than one-half of the free cruising distance shown in Table 3.1. Our analyses show a requirement of 29,221 total annual local sorties, all of which can be flown in areas within the standard distance. However, as discussed below, some of these areas do not meet requirements for size or other characteristics.

Ranges for Fighter Air-to-Ground Sorties

For most fighter air-to-ground sorties, local availability within standard distances must be determined using the relationships illustrated in Figure 2.3. This figure illustrates that the maximum free cruising distance must be equal to or greater than the sum of the distances from base to low-level route, low-level route to maneuver area surrounding a range, and range to base. An exception is made for A-10 sorties, in which low-level navigation tasks are assumed to be performed on random legs within a maneuver area rather than on a low-level route. For these A-10 sorties, the maneuver area associated with a range must be no more than one-half the free cruising distance from the base. Table 3.3 shows the specific low-level routes and maneuver areas used in this analysis.

Table 3.3

Cruising Distances Between Bases, Low-Level Routes, and Maneuver Areas (in nm)

Base	MDS	Route	Length of Route[a]	Distance (Base to Route)	Maneuver Area	Distance (Route to Area)	Distance (Range to Base)	Total Cruising Distance
Cannon	F-16CG, F-16GP	VR125	171	25	Pecos MOA	44	20	89
Davis-Monthan	A/OA-10[b]				Sells MOA	40	40	80
Hill	F-16CG	IR418 & Sevier MOA[c]	160	12	UTTR	0	12	24
Moody	A/OA-10[b]				Moody MOA	0	0	0
	F-16CG	VR1002	167	63	Moody MOA	47	0	110
Mt. Home	F-15E, F-16CJ	IR305	166	56	Paradise MOA	0	21	77
Pope	A/OA-10[b]				Poinsett MOA	108	108	216
Seymour-Johnson	F-15E	IR012	140	61	Dare County	0	85	146
Shaw	F-16CJ	VR087	167	63	Poinsett MOA	20	16	99

[a]Minimum required distance on a route is 160 nm. At an assumed speed of 480 kt, this distance allows 20 minutes of low-level navigation. Available lengths are determined by established MTR entry and exit points.

[b]A-10 sorties do not require a low-level route. In constructing sortie requirements embedded in the range and airspace database, PAF and ACC representatives determined that low-level navigation events in A-10 sorties are accomplished more effectively on random legs within a maneuver area than on a low-level route.

[c]IR418 and other low-level routes in the vicinity of Hill AFB are generally much shorter than 160 nm. However, they can be combined with the very large MOAs associated with the Utah Test and Training Range (UTTR) to provide a low-level navigation event of sufficient length.

Table 3.4 shows, for fighter bases, the number of local air-to-ground sorties to be flown annually and the number for which there is an available route and range configuration within the maximum free cruising distance. The data show that 81 percent of total annual sorties can be flown on ranges within the standard distance. The remaining sorties exceed the maximum free cruising distance, indicating that crews are receiving less than the standard duration of training, with a corresponding reduction in training value. Sorties exceeding the maximum free cruising distance occur at two bases—Pope and Moody AFBs. At Pope, the closest range (Poinsett) is too distant to permit any air-to-ground A-10 sorties within the maximum free cruising distance. At Moody AFB, as can be observed in Table 3.3, the distance to the closest available low-level route plus the distance from the route to the MOA above the Moody range exceeds the maximum free cruising distance for those F-16 air-to-ground sorties that require a low-level route.

Ranges for Bomber Sorties

Most bomber sorties are either CSS or SAT sorties with simulated delivery of weapons. Neither of these sorties requires a range upon which to drop ordnance.[1] However, given the increasingly important role for bombers in the delivery of conventional weapons, occasional access to an air-to-ground range is desirable. Several bomber bases (Barksdale, Ellsworth, and Minot) have no convenient access to such an asset, as indicated in Table 3.5.

ASSET QUALITY

Maneuver Areas

In this analysis, we evaluated the quality of the features present in the maneuver areas most commonly used by each base for its air-to-air and air-to-ground sorties. The analysis required the lateral

---

[1]Simulated delivery of ordnance requires an electronic scoring range, but information on electronic scoring ranges is not currently available in the range and airspace database. Thus, we were unable to evaluate these assets.

Table 3.4

Annual Sorties on Routes and Ranges Within Maximum Free Cruising Distances (MFCDs) (Fighter Air-to-Ground [A/G] Sorties)

Base	Sorties	A-10	OA-10	F-15E	F-16CG	F-16CJ	F-16GP	Total
Cannon	Local A/G				1,311		1,796	3,107
	within MFCD				1,311		1,796	3,107
	% within MFCD				100%		100%	100%
Davis-	Local A/G	2,513	817					3,330
Monthan	within MFCD	2,513	817					3,330
	% within MFCD	100%	100%					100%
Hill	Local A/G				7,657			7,657
	within MFCD				7,657			7,657
	% within MFCD				100%			100%
Moody	Local A/G	1,737	1,427		3,073			6,236
	within MFCD	1,737	1,427		1,622			4,786
	% within MFCD	100%	100%		53%			77%
Mt. Home	Local A/G			1,804		1,242		3,046
	within MFCD			1,804		1,242		3,046
	% within MFCD			100%		100%		100%
Pope	Local A/G	3,733	2,140					5,873
	within MFCD	0	0					0
	% within MFCD	0%	0%					0%
Seymour-	Local A/G			5,092				5,092
Johnson	within MFCD			5,092				5,092
	% within MFCD			100%				100%
Shaw	Local A/G					5,004		5,004
	within MFCD					5,004		5,004
	% within MFCD					100%		100%
Total	Local A/G	7,983	4,383	6,896	12,041	6,247	1,796	39,346
	within MFCD	4,250	2,243	6,896	10,590	6,247	1,796	32,023
	% within MFCD	53%	51%	100%	88%	100%	100%	81%

and/or vertical combination of adjacent MOAs, warning areas, and/or restricted areas into composites. These composites provide a block of airspace that can be compared with the dimensions specified in the range and airspace database for various MDS/sortie-type combinations. For example, to obtain the contiguous altitude required for many SAT sorties (300 ft to 25,000 ft), it may be necessary to combine a low MOA (100 ft to 8,000 ft), a high MOA (8,000 ft to 18,000 ft), and an ATCAA above the high MOA. Similarly, to provide sufficient lateral dimensions, it may be necessary to combine several adjacent MOAs or warning areas.

Table 3.5

Bomber Base Proximity to Air-to-Ground Ranges

Bomber Base	Nearest Air-to-Ground Range	Distance (nm)
Barksdale	Melrose	687
Dyess	Melrose	301
Ellsworth	UTTR	637
Minot	UTTR	895
Mt. Home	Saylor Creek	30
Whiteman	Smoky Hill	200

Table 3.6 contains data on the maneuver area characteristics in which deficiencies were noted for fighters. (We observed no deficiencies in any MDS/sortie-type combination for the following maneuver area characteristics: *over land, over mountains, over water, adjoining orbit,* and *adjoining range.* Accordingly, these characteristics do not appear in the table.) The table reveals that almost half of fighter sorties are flown in maneuver areas with insufficient lateral dimensions. Large proportions of sorties are also flown without required floors or ceilings. Chaff and flares are required but not authorized for about one-third of the sorties. Air combat maneuvering instrumentation (ACMI), datalink frequencies, and radar-jamming capabilities are generally unavailable.

Table 3.6

Annual Sorties by Maneuver Area Characteristics

Area Characteristic	Availability[a]	A-10	OA-10	F-15C	F-15E	F-16CG	F-16CJ	F-16GP	Total
Lateral dimensions[b]	Yes	2,834	966	8,984	5,761	7,939	3,862	2,490	32,835
		34%	21%	71%	60%	63%	38%	66%	53%
	No	5,470	3,566	3,663	3,858	4,758	6,197	1,301	28,813
		66%	79%	29%	40%	37%	62%	34%	47%
Floor[c]	Yes	4,380	2,488	10,779	3,813	7,653	9,013	694	38,820
		53%	55%	85%	40%	60%	90%	18%	63%
	No	3,923	2,044	1,868	5,806	5,044	1,046	3,097	22,828
		47%	45%	15%	60%	40%	10%	82%	37%
Ceiling[d]	Yes	4,013	2,065	12,647	4,527	8,282	2,288		33,822
		48%	46%	100%	47%	65%	23%		55%
	No	4,290	2,467		5,092	4,415	7,771		27,826
		52%	54%		53%	35%	77%		45%
Chaff authorized	Yes	2,024	1,656	10,788	1,206	6,266	3,660		25,599
		24%	37%	85%	13%	49%	36%		42%
	No	6,280	2,876	1,859	714	4,021	4,647	3,120	23,517
		76%	63%	15%	7%	32%	46%	82%	38%
	Not req'd				2,287	2,410	1,751	671	7,119
					24%	19%	17%	18%	12%
	Infrastructure unknown				5,412				5,412
					56%				9%
Flares authorized	Yes	2,407	2,065	10,788	1,206	6,266	3,660		26,392
		29%	46%	85%	13%	49%	36%		43%
	No	5,897	2,467	1,859	714	4,021	4,647	3,120	22,725
		71%	54%	15%	7%	32%	46%	82%	37%
	Not req'd				2,287	2,410	1,751	671	7,119
					24%	19%	17%	18%	12%
	Infrastructure unknown				5,412				5,412
					56%				9%
Supersonic authorized	Yes			7,414		855	2,625	1,583	12,476
				59%		7%	26%	42%	20%
	No			1,285	584	1,471	797		4,138
				10%	6%	12%	8%		7%
	Not req'd	8,304	4,532	3,948	7,389	8,323	6,637	2,208	41,341
		100%	100%	31%	77%	66%	66%	58%	67%
	Infrastructure unknown				1,646	2,047			3,693
					17%	16%			6%

Table 3.6 —continued

Area Characteristic	Availability[a]	A-10	OA-10	F-15C	F-15E	F-16CG	F-16CJ	F-16GP	Total
ACMI	No			8,761 69%	1,790 19%	4,617 36%	8,094 80%	2,708 71%	25,970 42%
	Not req'd	8,304 100%	4,532 100%	3,886 31%	2,780 29%	4,017 32%	1,965 20%	1,083 29%	26,566 43%
	Infrastructure unknown				5,049 52%	4,063 32%			9,112 15%
Air-to-air frequency	Yes	1,212 15%	670 15%	12,638 100%	714 7%	1,902 15%	3,636 36%	2,328 61%	23,099 37%
	No	16 0%	13 0%			2,012 16%	499 5%		2,540 4%
	Not req'd	7,076 85%	3,849 85%		4,719 49%	5,339 42%	5,800 58%	1,463 39%	28,246 46%
	Infrastructure unknown			9 0%	4,186 44%	3,445 27%	124 1%		7,763 13%
Air-to-ground frequency	Yes	7,076 85%	3,849 85%	25 0%	0%	3,573 28%	5,146 51%	1,796 47%	21,464 35%
	No			37 0%	1,804 19%		1,277 13%		3,119 5%
	Not req'd	1,227 15%	683 15%	12,585 100%	2,723 28%	5,980 47%	3,636 36%	1,995 53%	28,829 47%
	Infrastructure unknown				5,092 53%	3,144 25%			8,236 13%
Datalink frequency	Yes			2,700 21%					2,700 4%
	No			4,050 32%	1,661 17%	2,832 22%	7,948 79%	2,426 64%	18,917 31%
	Not req'd	8,304 100%	4,532 100%	5,897 47%	3,273 34%	5,624 44%	2,111 21%	1,364 36%	31,105 50%
	Infrastructure unknown				4,685 49%	4,241 33%			8,926 14%
Number of threat emitters	None	1,872 23%	934 21%	28 0%		1,451 11%	177 2%		4,460 7%
	Equals/ exceeds req'ment	2,441 29%	1,400 31%	9 0%	1,206 13%	2,016 16%	4,495 45%		11,567 19%

Table 3.6 —continued

Area Characteristic	Availability[a]	A-10	OA-10	F-15C	F-15E	F-16CG	F-16CJ	F-16GP	Total
Number of threat emitters (cont)	Not req'd	3,991 48%	2,198 48%	12,229 97%	4,676 49%	8,193 65%	5,263 52%	2,618 69%	39,168 64%
	Infrastructure unknown			380 3%	3,737 39%	1,037 8%	124 1%	1,173 31%	6,452 10%
Former Soviet Union (FSU) point emitter	Yes	2,441 29%	1,400 31%	176 1%	1,206 13%	2,108 17%	4,495 45%		11,826 19%
	No	1,872 23%	934 21%	241 2%	87 1%	1,517 12%	301 3%		4,952 8%
	Not req'd	3,991 48%	2,198 48%	12,229 97%	4,676 49%	8,193 65%	5,263 52%	2,618 69%	39,168 64%
	Infrastructure unknown				3,650 38%	879 7%		1,173 31%	5,702 9%
FSU area emitter	Yes	2,441 29%	1,400 31%	167 1%		2,108 17%	3,601 36%		9,717 16%
	No	1,872 23%	934 21%	241 2%	87 1%	1,517 12%	301 3%		4,952 8%
	Not req'd	3,991 48%	2,198 48%	12,229 97%	4,676 49%	8,193 65%	5,263 52%	2,618 69%	39,168 64%
	Infrastructure unknown			9 0%	4,856 50%	879 7%	894 9%	1,173 31%	7,810 13%
Non-FSU emitter	Yes	2,441 29%	1,400 31%	167 1%		2,108 17%	3,601 36%		9,717 16%
	No	1,872 23%	934 21%	250 2%	1,293 13%	1,517 12%	1,195 12%	0%	7,061 11%
	Not req'd	3,991 48%	2,198 48%	12,229 97%	4,676 49%	8,193 65%	5,263 52%	2,618 69%	39,168 64%
	Infrastructure unknown				3,650 38%	879 7%		1,173 31%	5,702 9%
Mobile emitter	Yes	2,441 29%	1,400 31%	176 1%	1,206 13%	2,108 17%	4,495 45%		11,826 19%
	No	1,872 23%	934 21%	241 2%	87 1%	1,517 12%	301 3%		4,952 8%
	Not req'd	3,991 48%	2,198 48%	12,229 97%	4,676 49%	8,193 65%	5,263 52%	2,618 69%	39,168 64%
	Infrastructure unknown				3,650 38%	879 7%		1,173 31%	5,702 9%

Table 3.6 —continued

Area Charac-teristic	Avail-ability[a]	A-10	OA-10	F-15C	F-15E	F-16CG	F-16CJ	F-16GP	Total
Debrief capa-bility	Yes	2,441	1,400			2,108	3,601		9,551
		29%	31%			17%	36%		15%
	No	1,872	934	417	1,293	1,517	1,195		7,227
		23%	21%	3%	13%	12%	12%		12%
	Not req'd	3,991	2,198	12,229	4,676	8,193	5,263	2,618	39,168
		48%	48%	97%	49%	65%	52%	69%	64%
	Infra-structure unknown				3,650	879		1,173	5,702
					38%	7%		31%	9%
Reactive emitter	Yes	2,441	1,400			2,108	3,601		9,551
		29%	31%			17%	36%		15%
	No	1,872	934	417	1,293	1,517	1,195		7,227
		23%	21%	3%	13%	12%	12%		12%
	Not req'd	3,991	2,198	12,229	4,676	8,193	5,263	2,618	39,168
		48%	48%	97%	49%	65%	52%	69%	64%
	Infra-structure unknwon				3,650	879		1,173	5,702
					38%	7%		31%	9%
Smokey SAMs	Yes	736				2,016			2,751
		9%				16%			4%
	No	3,577	2,334	62	1,206	1,451	4,672		13,301
		43%	52%	0%	13%	11%	46%		22%
	Not req'd	3,991	2,198	12,585	5,010	8,390	5,387	2,665	40,226
		48%	48%	100%	52%	66%	54%	70%	65%
	Infra-structure unknown				3,403	840		1,125	5,369
					35%	7%		30%	9%
Radar jam-ming	No	4,313	2,334	417	1,293	3,625	4,796		16,778
		52%	52%	3%	13%	29%	48%		27%
	Not req'd	3,991	2,198	12,229	4,676	8,193	5,263	2,618	39,168
		48%	48%	97%	49%	65%	52%	69%	64%
	Infra-structure unknown				3,650	879		1,173	5,702
					38%	7%		31%	9%

Table 3.6 —continued

Area Characteristic	Availability[a]	A-10	OA-10	F-15C	F-15E	F-16CG	F-16CJ	F-16GP	Total
Communications jamming	Yes					92			92
						1%			0%
	No			355	87	66	124		633
				3%	1%	1%	1%		1%
	Not req'd	8,304	4,532	12,291	9,285	12,500	9,935	3,743	60,590
		100%	100%	97%	97%	98%	99%	99%	98%
	Infrastructure unknown				247	38		48	333
					3%	0%		1%	1%

[a]"Yes" indicates that the characteristic is required and available. "No" indicates that the characteristic is required and not available. "Infrastructure unknown" indicates that information on the airspace infrastructure is missing in the range and airspace database.

[b]Indicates whether the area has the required length and width.

[c]Indicates whether the floor of the area is low enough to meet requirements.

[d]Indicates whether the ceiling of the area is high enough to meet requirements.

Low-Level Routes

Table 3.7 shows the number of fighter sorties that must be flown on routes with deficient characteristics. The only route characteristics on which we did not note deficiencies were *segment below 300 feet* (available on all routes), *access to special use airspace* (SUA) (available on all routes), and *communications jamming* (not required for any sortie type); accordingly, these characteristics were excluded from the table. The table reveals that over half the routes lack required width and almost a third lack required length. Required floors, terrain-following flight, and 5000-ft segments are required but unavailable on about a third of the sorties. Very few sorties are flown over mountainous terrain. Most routes lack threat emitters, radar jamming, and debrief capability.

Table 3.7

Annual Sorties by Route Characteristics

Route Characteristic	Availability[a]	F-15E	F-16CG	F-16CJ	F-16GP	Total
Width	Yes		1,162	4,724	1,529	7,415
			37%	80%	100%	42%
	No	6,896	2,005	1,173		10,073
		100%	63%	20%		58%
Length	Yes	1,804	3,167	5,897	1,529	12,396
		26%	100%	100%	100%	71%
	No	5,092				5,092
		74%				29%
Floor	Yes	1,804	3,167	5,897	1,529	12,396
		26%	100%	100%	100%	71%
	No	5,092				5,092
		74%				29%
Ceiling	Yes	1,804	1,162	5,897	1,529	10,392
		26%	37%	100%	100%	59%
	No		2,005			2,005
			63%			11%
	Infrastructure unknown	5,092				5,092
		74%				29%
Terrain following	Yes	6,896	3,167	1,173	1,529	12,764
		100%	100%	20%	100%	73%
	No			4,724		4,724
				80%		27%
25 nm segment to 5000 ft	Yes	1,804				1,804
		26%				10%
	No	5,092				5,092
		74%				29%
	Not required		3,167	5,897	1,529	10,592
			100%	100%	100%	61%
50% mountainous	Yes	1,804		1,173		2,977
		26%		20%		17%
	No	5,092	3,167	4,724	1,529	14,511
		74%	100%	80%	100%	83%
Number of threat emitters	Meets requirement			3,321		3,321
				56%		19%
	None	4,609	1,194	824		6,627
		67%	38%	14%		38%
	Not required	2,287	1,282	1,751	671	5,991
		33%	40%	30%	44%	34%
	Infrastructure unknown		691		858	1,549
			22%		56%	9%

Table 3.7—continued

Route Characteristic	Availability[a]	F-15E	F-16CG	F-16CJ	F-16GP	Total
FSU point emitter	No	4,609	1,194	4,146		9,949
		67%	38%	70%		57%
	Not required	2,287	1,282	1,751	671	5,991
		33%	40%	30%	44%	34%
	Infrastructure unknown		691		858	1,549
			22%		56%	9%
FSU area emitter	No	4,609	1,194	4,146		9,949
		67%	38%	70%		57%
	Not required	2,287	1,282	1,751	671	5,991
		33%	40%	30%	44%	34%
	Infrastructure unknown		691		858	1,549
			22%		56%	9%
Non-FSU emitter	No	4,609	1,194	4,146		9,949
		67%	38%	70%		57%
	Not required	2,287	1,282	1,751	671	5,991
		33%	40%	30%	44%	34%
	Infrastructure unknown		691		858	1,549
			22%		56%	9%
Mobile emitter	Yes			3,321		3,321
				56%		19%
	No	4,609	1,194	824		6,627
		67%	38%	14%		38%
	Not required	2,287	1,282	1,751	671	5,991
		33%	40%	30%	44%	34%
	Infrastructure unknown		691		858	1,549
			22%		56%	9%
Reactive emitter	Yes			3,321		3,321
				56%		19%
	No	4,609	1,194	824		6,627
		67%	38%	14%		38%
	Not required	2,287	1,282	1,751	671	5,991
		33%	40%	30%	44%	34%
	Infrastructure unknown		691		858	1,549
			22%		56%	9%
Smoky SAMs	No	4,609	1,885	4,146	858	11,497
		67%	60%	70%	56%	66%
	Not required	2,287	1,282	1,751	671	5,991
		33%	40%	30%	44%	34%

Table 3.7—continued

Route Characteristic	Availability[a]	F-15E	F-16CG	F-16CJ	F-16GP	Total
Radar jamming	No	4,609	1,885	4,146	858	11,497
		67%	60%	70%	56%	66%
	Not required	2,287	1,282	1,751	671	5,991
		33%	40%	30%	44%	34%
Debrief capability	Yes			3,321		3,321
				56%		19%
	No	4,609	1,194	824		6,627
		67%	38%	14%		38%
	Not required	2,287	1,282	1,751	671	5,991
		33%	40%	30%	44%	34%
	Infrastructure		691		858	1,549
	unknown		22%		56%	9%

[a]"Yes"indicates that the characteristic is required and available. "No" indicates that the characteristic is required and not available. "Infrastructure unknown" indicates that information on the airspace infrastructure is missing in the range and airspace database.

Ranges

In this analysis, we evaluated the quality of the features on the primary range used by each base for its air-to-ground sorties. For bombers, we observed no deficiencies other than limitations on weapon deliveries imposed by size of the WSFA and restricted airspace. For fighters, we observed no deficiencies in any MDS/ sortie-type combination for the following range characteristics (in all cases, the characteristic is either available or not required): *conventional circles, strafable targets, 30-millimeter munitions capable targets, lighted targets, vertical targets, tactical target array, second tactical target array, urban target array, strafe scoring, night scoring, scoring within 15 seconds, classified operations, over water operations, communications jamming,* and *range control officer availability.* Table 3.8 contains data on binary (yes/no) characteristics in which deficiencies were noted. Total sorties affected by the deficiency are shown. Table 3.9 contains similar data on ordnance types allowed. Table 3.10 contains data for characteristics that are expressed as numerical quantities (e.g., if a minimum number of targets is specified for a sortie type, Table 3.10 indicates the number of sorties by percentage of the required number of targets that are available).

Table 3.8

Annual Sorties by Range Characteristics

Range Charac- teristic	Avail- ability[a]	A-10	OA-10	F-15E	F-16CG	F-16CJ	F-16GP	Total
Laser scoring	Yes				4,456			4,456
					66%			13%
	No		2,471	6,896	2,262			11,628
			56%	100%	34%			34%
	Not req'd	7,983	1,912			6,247	1,796	17,938
		100%	44%			100%	100%	53%
Score to 1 meter	Yes	6,246	2,957	6,896	4,456	6,247	1,796	28,597
		78%	67%	100%	66%	100%	100%	84%
	No	1,737	1,427		2,262			5,425
		22%	33%		34%			16%
	Not req'd							
Chaff authorized	Yes	1,642	534	4,609	2,856	894	1,125	11,661
		21%	12%	67%	43%	14%	63%	34%
	No	3,577	2,334		1,451	3,601		10,963
		45%	53%		22%	58%		32%
	Not req'd	2,764	1,515	2,287	2,411	1,751	671	11,398
		35%	35%	33%	36%	28%	37%	34%
Flares authorized	Yes	4,083	1,935	4,609	2,856	4,495	1,125	19,104
		51%	44%	67%	43%	72%	63%	56%
	No	1,136	934		1,451			3,520
		14%	21%		22%			10%
	Not req'd	2,764	1,515	2,287	2,411	1,751	671	11,398
		35%	35%	33%	36%	28%	37%	34%
ECM authorized	Yes	4,083	1,935	4,609	2,856	4,495	1,125	19,104
		51%	44%	67%	43%	72%	63%	56%
	No	1,136	934		1,451			3,520
		14%	21%		22%			10%
	Not req'd	2,764	1,515	2,287	2,411	1,751	671	11,398
		35%	35%	33%	36%	28%	37%	34%
Illumination flares authorized	Yes	6,246	1,290					7,536
		78%	29%					22%
	No	1,737	622					2,359
		22%	14%					7%
	Not req'd		2,471	6,896	6,718	6,247	1,796	24,127
			56%	100%	100%	100%	100%	71%
90-deg rotation of axis of attack available	Yes	1,642	534	4,609	2,856	894	1,125	11,661
		21%	12%	67%	43%	14%	63%	34%
	No	3,577	2,334		1,451	3,601		10,963
		45%	53%		22%	58%		32%
	Not req'd	2,764	1,515	2,287	2,411	1,751	671	11,398
		35%	35%	33%	36%	28%	37%	34%

Table 3.8—continued

Range Charac-teristic	Avail-ability[a]	A-10	OA-10	F-15E	F-16CG	F-16CJ	F-16GP	Total
Night-vision goggles (NVG) capable	Yes	6,246	2,957	5,092	4,456	5,004	1,796	25,551
		78%	67%	74%	66%	80%	100%	75%
	No	1,737	1,427	1,804	2,262	1,242		8,471
		22%	33%	26%	34%	20%		25%
	Not req'd							
FSU point emitter	Yes	2,441	1,400	4,609	2,856	4,495	1,125	16,927
		31%	32%	67%	43%	72%	63%	50%
	No	2,778	1,468		1,451			5,697
		35%	33%		22%			17%
	Not req'd	2,764	1,515	2,287	2,411	1,751	671	11,398
		35%	35%	33%	36%	28%	37%	34%
FSU area emitter	Yes	2,441	1,400	4,609	840	4,495	1,125	14,911
		31%	32%	67%	13%	72%	63%	44%
	No	2,778	1,468		3,466			7,713
		35%	33%		52%			23%
	Not req'd	2,764	1,515	2,287	2,411	1,751	671	11,398
		35%	35%	33%	36%	28%	37%	34%
Non-FSU emitter	Yes	2,441	1,400	3,403	840	3,601	1,125	12,812
		31%	32%	49%	13%	58%	63%	38%
	No	2,778	1,468	1,206	3,466	894		9,812
		35%	33%	17%	52%	14%		29%
	Not req'd	2,764	1,515	2,287	2,411	1,751	671	11,398
		35%	35%	33%	36%	28%	37%	34%
Mobile emitter	Yes	2,441	1,400	4,609	840	4,495	1,125	14,911
		31%	32%	67%	13%	72%	63%	44%
	No	2,778	1,468		3,466			7,713
		35%	33%		52%			23%
	Not req'd	2,764	1,515	2,287	2,411	1,751	671	11,398
		35%	35%	33%	36%	28%	37%	34%
Debrief capability	Yes	2,441	1,400	4,609	840	4,495	1,125	14,911
		31%	32%	67%	13%	72%	63%	44%
	No	2,778	1,468		3,466			7,713
		35%	33%		52%			23%
	Not req'd	2,764	1,515	2,287	2,411	1,751	671	11,398
		35%	35%	33%	36%	28%	37%	34%
Reactive emitter	Yes	2,441	1,400		840	3,601	1,125	9,408
		31%	32%		13%	58%	63%	28%
	No	2,778	1,468	4,609	3,466	894		13,216
		35%	33%	67%	52%	14%		39%
	Not req'd	2,764	1,515	2,287	2,411	1,751	671	11,398
		35%	35%	33%	36%	28%	37%	34%

Table 3.8—continued

Range Charac- teristic	Avail- ability[a]	A-10	OA-10	F-15E	F-16CG	F-16CJ	F-16GP	Total
Smokey SAMs	Yes	5,219	2,868	3,403	4,306	3,601	1,125	20,524
		65%	65%	49%	64%	58%	63%	60%
	No			1,206		894		2,100
				17%		14%		6%
	Not req'd	2,764	1,515	2,287	2,411	1,751	671	11,398
		35%	35%	33%	36%	28%	37%	34%
Radar jamming	Yes							
	No	5,219	2,868	4,609	4,306	4,495	1,125	22,624
		65%	65%	67%	64%	72%	63%	66%
	Not req'd	2,764	1,515	2,287	2,411	1,751	671	11,398
		35%	35%	33%	36%	28%	37%	34%

[a]"Yes" indicates that the characteristic is required and available. "No" indicates that the characteristic is required and not available.

Table 3.9

Annual Sorties by Range Ordnance Restrictions

Restriction	A-10	OA-10	F-15E	F-16CG	F-16CJ	F-16GP	Total
Live ordnance	26	6	256	105	18	48	458
required/not allowed	0%	0%	4%	2%	0%	3%	1%
Inert ordnance	1,711	1,420		3,325			6,456
required/not allowed	21%	32%		49%			19%
All required ordnance	6,246	2,957	6,640	3,288	6,229	1,749	27,108
types allowed	78%	67%	96%	49%	100%	97%	80%

Table 3.10

Annual Sorties on Ranges by Available Proportion of Required Targets, Scorers, or Threat Emitters

Range Characteristic	Proportion of Requirement Satisfied	A-10	OA-10	F-15E	F-16CG	F-16CJ	F-16GP	Total
Laser targets	0%	5,469	3,566		2,261	2,460		13,757
		69%	81%		34%	39%		40%
	100% or more			6,895	4,456	610	267	12,230
				100%	66%	10%	15%	36%
	Not req'd					3,175	1,528	4,704
						51%	85%	14%
	Missing data	2,513	816					3,330
		31%	19%					10%
Heated targets	25%				2,016			2,016
					30%			6%
	75%	2,441	1,400			3,601		7,443
		31%	32%			58%		22%
	100% or more	3,029	2,166	6,896	4,702	2,645	1,796	21,234
		38%	49%	100%	70%	42%	100%	62%
	Missing data	2,513	817					3,330
		31%	19%					10%
Radar targets	0%		2,010		2,262	5,004		9,276
			46%		34%	80%		27%
	100% or more	0		6,896	4,456	1,242	1,796	14,391
		0%		100%	66%	20%	100%	42%
	Not required	7,983	1,912					9,895
		100%	44%					29%
	Missing data		461					461
			11%					1%
Simultaneously scored targets	25%	3,577	2,334		1,451	3,601		10,963
		45%	53%		22%	58%		32%
	75%			1,206		894		2,100
		0%		17%		14%		6%
	100% or more	4,406	2,049	5,690	2,122	1,751	1,796	17,815
		55%	47%	83%	32%	28%	100%	52%
	Missing data				3,145			3,145
					47%			9%

Table 3.10—continued

Range Charac- teristic	Proportion of Requirement Satisfied	A-10	OA-10	F-15E	F-16CG	F-16CJ	F-16GP	Total
Threat emit- ters	0%	2,778	1,468		1,451			5,697
		35%	33%		22%			17%
	50%				2,016			2,016
					30%			6%
	100% or more	2,441	1,400	4,609	840	4,495	1,125	14,911
		31%	32%	67%	13%	72%	63%	44%
	Not required	2,764	1,515	2,287	2,411	1,751	671	11,398
		35%	35%	33%	36%	28%	37%	34%

Predominant problems noted in these tables include lack of laser scoring, chaff authorization, multiple axes of attack, use of night vision goggles, threat emitter variety, radar jamming, debrief capability, and laser and radar targets.

Weapon Deliveries

WSFA and restricted airspace sizes place limits on the types of weapon deliveries that are permitted on a given range. The range and airspace database allows users to generate lists of allowable deliveries on a given range. As of August 1999, ACC/DOR had identified weapon safety footprints for 210 distinct delivery types (MDS, delivery mode, weapon combinations). Table 3.11 indicates, for the ranges most commonly used by ACC aircrews, how many of these delivery types could be accommodated, assuming a 2 nm × 2 nm tactical target array.[2]

---

[2]Some ranges may support more than the number of delivery types indicated in Table 3.11 by providing a target array smaller than the standard 2 nm × 2 nm specified here or by restricting which targets may be used for certain deliveries. Additionally, some ranges may support less than the number shown in Table 3.11 because of unfavorable placement of the target array relative to the boundaries of the WSFA. Our count of accommodated delivery types is premised on optimal placement of the target array within the available WSFA.

Table 3.11

Number of Weapon Delivery Types Accommodated on Commonly Used Ranges (n = 210)

Range	Accommodated	Not Accommodated	Unknown Infra- structure[a]	Unknown Require- ment[a]
Restricted airspace, lateral dimensions				
Dare County	202			8
Goldwater	202			8
Grand Bay	199	3		8
Melrose	202			8
Poinsett	173	29		8
Saylor Creek	202			8
Townsend	202			8
UTTR				
R6402	202			8
R6404	202			8
R6405	202			8
R6406	202			8
R6407	202			8
Restricted airspace, ceiling				
Dare County	173	31		6
Goldwater			210	6
Grand Bay	173	31		6
Melrose	173	31		6
Poinsett	173	31		6
Saylor Creek	166	38		6
Townsend	149	55		6
UTTR				
R6402	204			6
R6404	204			6
R6405	204			6
R6406	204			6
R6407	204			6

Table 3.11—continued

Range		Accommodated	Not Accommodated	Unknown Infra-structure[a]	Unknown Require-ment[a]
WSFA					
	Dare County		210		
	Goldwater	210			
	Grand Bay	27	183		
	Melrose	146	64		
	Poinsett		210		
	Saylor Creek	173	37		
	Townsend	106	104		
	UTTR				
	R6402	210			
	R6404	210			
	R6405			210	
	R6406	210			
	R6407	210			

[a]"Unknown infrastructure" indicates that information needed to evaluate the range is unknown. "Unknown requirement" indicates that weapon delivery information needed to determine the required dimension is unknown.

ASSET CAPACITY

Maneuver Areas and Low-Level Routes

We examined both fighter and bomber access to maneuver areas and low-level routes for both air-to-air and air-to-ground sortie requirements, and found that no base currently faces a capacity constraint on either type of asset.

Ranges

Table 3.12 shows that all fighter bases but one have access to sufficient range capacity to meet their annual air-to-ground sortie requirements, although not necessarily on their own ranges. ACC aircrews at Davis-Monthan AFB reportedly have access to the

Table 3.12

Fighter Air-to-Ground Range Requirements Versus Capacities

Base	Annual Hours Required	Range	Primary Asset Annual Hours Available	Range	Secondary Asset Annual Hours Available
Cannon	1,027	Melrose	3,211		
Davis-Monthan	1,611	Goldwater	1,085		
Hill	1,008	UTTR	5,039		
Moody	3,674	Grand Bay	3,497	Townsend	1,679
Mt. Home	831	Saylor Creek	3,786		
Pope	3,079	Poinsett	3,489	Dare County	5,897
Seymour-Johnson	1,391	Dare County	5,897		
Shaw	1,362	Poinsett	3,489		

Goldwater range for only 25 percent of its operating hours. If this estimate is correct, crews at Davis-Monthan do not have sufficient range time to meet annual requirements. Moody has slightly more requirements than can be accommodated on Grand Bay, its adjacent range, and must complete some requirements at the more distant Townsend range operated by the Georgia Air National Guard. Finally, Pope aircrews require more time than is available on the Poinsett range (after Poinsett satisfies Shaw AFB requirements), requiring Pope aircrews to use the more distant Dare County range for some of their requirements.

Relative to fighters, bombers have fewer requirements for access to air-to-ground ranges and are able to reach ranges at much greater distances from their home bases. Traditional bomber training has taken place on low-level training routes with radar bomb-scoring sites, which do not require the crew to actually release a weapon from the aircraft. This is changing as the bomber community adjusts to new weapons and a greater emphasis on precision delivery of Joint Direct Attack Munitions (JDAMs) and standoff delivery capability. Clearly, some regular local access to a range complex would result in lower unit sortie duration times and more efficient training programs. As indicated in Table 3.5, distances to air-to-ground ranges may mean lengthy crew duty days and much nontraining flight time en route to ranges for some bomber crews. However, because their aircraft are capable of lengthy sorties, bomber crews do not face any range capacity constraints given the current small number of RAP sorties requiring actual releases.[3]

SUMMARY

In this chapter, we assessed the range and airspace assets used by ACC units in three ways—geographically, qualitatively, and quantitatively. We noted problems in each of these assessments.

We found that proximity to assets is a problem only for air-to-ground sorties. For A-10s at Pope AFB and F-16s at Moody AFB, aircrews get less actual training time than their counterparts at other bases because of geographical separation from their training assets. For bombers, crews at some bases have to cover inordinate distances to experience weapon releases.

Qualitatively, large proportions of fighter sorties are flown using routes, areas, and ranges with substandard characteristics. Collectively, these deficiencies make it difficult for crews to experience, and learn to react to, the threats and conditions they must be prepared to encounter in combat.

Quantitatively, we found sufficient capacity on routes and areas but some limitations on ranges. One base, Davis-Monthan, appears to face insufficient capacity, while two others (Pope and Moody) must split their sorties between two ranges to meet their capacity requirements.

---

[3]As it is presently configured, RAP allows a wide range of training programs in bomber units. We found that units with much better access to local ranges (Whiteman and Mountain Home AFBs) were much more likely to use those ranges in their training programs. Other units, without a nearby range, were more likely to use radar scoring sites and low-level routes to which they had much better access. One unit, located at Dyess AFB, is trying to lead a major realignment of training routes, MOAs, and ranges in Texas, Colorado, and eastern New Mexico to allow it opportunity to conduct more effective and efficient training. Bombers without a low-level requirement (B-2s at Whiteman, B-52s at Barksdale AFB) have more flexibility in meeting their training requirements except those few events relating to high-altitude release of inert or live weapons.

Chapter Four

ONGOING ANALYSIS CAPABILITIES AND APPLICATIONS

In this chapter, we describe how the range and airspace database, used for the analyses reported in Chapter Three and subsequently delivered to ACC/DOR, can support ongoing staff processes.

EXAMINING INFRASTRUCTURE CHARACTERISTICS

The approach we took to examining infrastructure characteristics in our analysis was to determine total required annual sorties and then observe the proportion of the total in which some infrastructure characteristic is deficient. The results can be aggregated by base, MDS, sortie type, or any combination of these.[1] For example, data in Table 3.6 indicate that 47 percent of all air-to-air fighter sorties are flown in maneuver areas that are too small in their lateral dimensions. The problem is most severe for OA-10 units, which fly 79 percent of their sorties in maneuver areas that are too small.

Used in this way, the database provides part of the information that range and airspace managers need to evaluate how range and airspace infrastructure affects training. Specifically, it allows managers to depict the pervasiveness of various deficiencies. To fully evaluate the impact of these deficiencies, managers must supply their own sense of how seriously a given deficiency degrades training.

---

[1]With the addition to the database of a table specifying assignment of units to air expeditionary forces (AEFs), results could also be aggregated by AEF.

EXAMINING INFRASTRUCTURE CAPACITIES

An important feature of the database is that it generates a count of required annual sorties by base, MDS, and sortie type. Other information, such as the time required on a range or in a maneuver area for each sortie type, is combined with the sortie count to determine the annual hours each base needs on ranges and in various kinds of airspace. By comparing this need with the operating hours of ranges and airspace in proximity to each base, managers can determine whether sufficient capacity exists. If pressed to reduce the supply of ranges and airspace, managers can use the database to determine which assets are least critical, given their distances from using bases and the demands generated by the missions at those bases.

EVALUATING RESOURCING AND INVESTMENT OPTIONS

The database provides information for range and airspace managers to use in evaluating the impacts of alternative investments in range and airspace equipment or facilities. A potential investment that would remove a deficiency on a given asset can be evaluated in terms of how many sorties would be improved. Costs of alternatives can be divided by the number of improved sorties to determine which investments are most cost-effective.

Such an analysis has its limitations. It can be used to decide which *location* would provide the most cost-effective investment to reduce a specific deficiency, such as lack of a specific type of threat emitter. The database does not shed much light on the question of which *deficiency* should be treated first. For example, it does little to help evaluate the criticality of a deficiency related to threat emitters versus one related to communications or radar jammers. Managers must supply their own judgment regarding the training value of eliminating various deficiencies.

EVALUATING BASING OPTIONS

Some installations have better access than others to ranges and airspace. This access, like other features of an installation, should be weighed in any unit realignment or base closure decision. The range and airspace database allows decisionmakers to systematically review their options. Tables representing alternative unit beddowns can be created within the range and airspace database. The database's assessment features can then be exercised to quantify the proportion of sorties effectively supported by various beddown alternatives.

MANAGING AIRCREW TRAINING

The range and airspace database can be usefully extended to enhance the development, specification, and evaluation of aircrew training programs. Using the joint mission framework as a statement of operational needs, it should be possible to design applied RAP training sorties to meet future needs of joint force commanders. Additionally, it is possible to use the database in the production of Air Force AFI-11 series instructions, Standard Training Plans (STP), and other staff products disseminating training information.

Building a Training Program Based on Joint Need for Effects

One of the initial challenges we faced was the need to link aircrew training directly to the needs of joint force commanders. We accomplished that in our analytic structure by extending linkages backward from existing RAP sortie definitions to a joint mission framework. To a very limited degree, we created new sortie definitions to cover some apparent training needs, involving multiple MDS, not currently recognized by RAP.[2] Our creation of these new sortie types is an indication that RAP does not comprehensively meet the needs of joint force commanders. Reversing our approach—starting with the joint mission framework and determining the kinds of training sorties needed to support it—would provide a more systematic assessment and specification of training requirements.

Using the database in this way would require further development of the joint mission framework. The framework provides an inventory of needed operational effects. To be used as a guide for aircrew training programs, the subset of operational effects to which airpower can make a significant contribution would have to be

_____
[2] We refer to these as SMMEs.

identified. Within that subset, needed operational effects would have to be prioritized so that related training requirements could be given appropriate emphasis. Additionally, development of doctrine and concepts of operation might be required as intervening steps between identification of the needed operational effects and specification of supporting training programs.

Updating Training-Related Plans and Directives

ACC is responsible for developing a number of aircrew training-related documents at an MDS level of detail. These include standard training plans, AFI 11-2 series publications, and annual RAP tasking messages. Each of these documents contains an extensive amount of tabular information.

Embedding this tabular information in a relational database (ideally, an extension of the range and airspace database) would have several advantages. It would make the information readily available in electronic form for other applications, such as programming for resources needed to support training. It would lead to standardization of terms used across MDS, increasing the coherence of the entire aircrew training system and facilitating the development and specification of multiple-MDS training requirements. It would permit update of the information on a more frequent basis than current publication cycles. To the extent that conventional publications continue to have a role, standard reports derived from the database could be used to facilitate updates.

Chapter Five

OBSERVATIONS AND CONCLUSIONS

We next provide our broad observations on all elements of the analytic structure—operational requirements, training requirements, infrastructure requirements, and current infrastructure.

OPERATIONAL REQUIREMENT FRAMEWORKS

Seeking a strategies-to-task framework that would allow us to depict the linkages between training requirements and operational requirements, we found the available frameworks unacceptable. CINCs' war plans and unit DOCs are too detailed and context specific. Additionally, their level of classification makes their use impractical for a system designed for open communication with the public. The UJTL and its derivatives, the JMETL and AFTL, suffer from a land-centric orientation and a failure to recognize the contributions of aerospace power at strategic and operational levels of war. Consequently, we linked training requirements to our own statement of operational requirements—the *joint mission framework*. We believe this framework provides a comprehensive catalog of CINCs' operational needs, in both warfighting and peacetime employments, at three levels (mission, objective, and task). It should be useful in any study or analysis that requires component capabilities to be linked to joint warfighting requirements.

This joint mission framework is consistent with *Joint Vision 2010* (Chairman of the Joint Chiefs of Staff, undated) force development, an emerging Air Force emphasis on effects-based operations, and the concept of CINCs expressing requirements for generic aerospace force capabilities rather than specific weapon systems. It does not replace *Joint Vision 2010,* DOCs, or Air Force core competencies, but it complements them by allowing linkages between operational and training requirements to be clearly depicted.

TRAINING REQUIREMENTS

We found that, in many respects, aircrew training requirements are not formally specified at sufficient detail to derive requirements for range and airspace infrastructure or other training resources. Each MDS has a lengthy aircrew training directive (Volume 1 of the MDS-specific AFI 11-2 series), supplemented by annual RAP tasking messages. These documents provide detailed procedures for counting sortie types and training events but are often vague regarding the content and context of a sortie. Some specific deficiencies are noted below.

Duration of Training Events

In our interviews with aircrews in operational units, we noted that there is a wide range of practices, and no recognized standard, for how much of a sortie should be dedicated to specified training events as opposed to cruising to and from training areas. For example, SAT missions require a low-level navigation event and a weapon delivery event. Neither the minimum time or distance to be covered in the navigation event nor the number of weapon deliveries, or time spent on weapon delivery events, is specified. Units that routinely have their crews do 10 minutes of low-level navigation and two weapon delivery passes receive roughly half the training value on a sortie as units that have their crews do 20 minutes of low-level navigation and four passes. To establish reasonable standards for geographical proximity of ranges and airspace to bases and to compute total expected asset usage for the purpose of examining range and airspace capacities, we had to establish our own tentative standards for durations of training events.

Threats

Various air-to-ground sorties require execution of training events in an environment that includes actual or simulated threats. Threats, presumably, may be ground-to-air or air-to-air. The nature of the threats is unspecified in training documentation. The need for ground-to-air threats is implied by the chaff and flare events included in annual RAP tasking messages, but we found no mention in training documentation of a requirement that some sorties be air-opposed. If air-opposed sorties are not required, maneuver areas surrounding a range can be relatively small on both lateral and vertical dimensions. We assumed that some air-opposed ground-attack sorties are required and constructed maneuver area requirements accordingly.

Multi-MDS Requirements

With the exception of FAC-A and LFE sorties and refueling events, we found no requirements for training events involving multiple MDS. In actual employments, aircraft in surface attack and counter-air roles must often interact with each other, and both must interact with C^2ISR assets (ground or airborne). There are no documented requirements for small-scale, building-block exercises (involving, for example, a flight of fighters and a JSTARS or AWACS platform or a flight of fighters or bombers and supporting SEAD aircraft) to build the necessary coordination skills. However, we noted some units in the field seeking to arrange such sorties and small-scale exercises on their own initiative. Our analysis indicates that such exercises would require larger maneuver areas than single-MDS ground-attack sorties. We postulated the need for such sorties and identified several examples.

INFRASTRUCTURE REQUIREMENTS

Prior to our analysis, infrastructure requirements were specified primarily in relation to safety considerations. Their relationships to training requirements were vague and implicit. An important contribution of the range and airspace database is that it makes the links between infrastructure and training requirements explicit and demonstrable. As ACC/DOR gains experience with the infrastructure requirements currently captured in the database, opportunities for useful refinement and expansion of the database will likely emerge. For example, ACC/DOR has recently indicated a need for greater detail regarding threat emitter requirements.

CURRENT INFRASTRUCTURE

When PAF began this study, centralized repositories of information on ranges and airspace were very limited. For ranges, there was a partially outdated database assembled by a contractor for an Air Staff client, but access to the database was limited and there were no provisions for updating the data. For airspace, NIMA's DAFIF contains some geographical data but lacks information on other characteristics needed to compare available airspace with training-related requirements.

The range and airspace database partially fills this gap. As currently configured, it serves as a repository for training-related infrastructure requirements and training-related characteristics of current infrastructure. It also provides rudimentary capabilities for updates from either headquarters or field sources. Building on this foundation, ACC/DOR has the opportunity to expand the database to include other range- and airspace-related management information. For example, the database could be used to capture range and airspace utilization data, providing ACC/DOR a cross-check against the capacity demands now computed and recorded in the database. Other useful expansions would include inventories of targets, threat emitters, scoring systems, and other installed equipment; requirements and infrastructure from non-ACC range and airspace users; and information regarding noninfrastructure training resource demands (e.g., flying hours, munitions, and maintenance effort).

INFRASTRUCTURE ASSESSMENTS

From the assessments provided in Chapter Three, we can draw some general conclusions regarding current infrastructure.

Proximity

We found no proximity problems for air-to-air sorties but did find some for air-to-ground sorties. Pope AFB, lacking a backyard range, is too distant from Dare County and Poinsett ranges to afford minimum standard training event durations on any of its air-to-ground sorties. Moody AFB has a range immediately adjacent to the base, but Moody F-16s cannot meet event duration standards for applied air-to-ground sorties because of distances to and from the nearest low-level route. Additionally, crews from several bomber bases (Barksdale, Ellsworth, and Minot) must travel a very long distance to actually drop a bomb or launch a standoff weapon.

Characteristics

Insufficient lateral or vertical dimensions are a problem for a large proportion of MOAs, MTRs, and WSFAs. Deficiencies are widely observed in chaff and flare use authorizations, scoring and other feedback systems, ordnance types permitted, targets, threat emitters, terrain variety, and other characteristics.

Capacity

Capacity is generally not a problem. Only Davis-Monthan AFB appears to face a capacity constraint. Pope has no backyard range and Moody has insufficient capacity on its backyard range, but both can obtain needed capacity by traveling to more distant ranges (at the cost of reducing training event durations to less than standardized minimums).

KEEPING THE DATABASE VIABLE

The range and airspace database provides a powerful tool for range and airspace managers and a potential tool for other aircrew training resource managers. To remain viable, it must be maintained and updated, which will require an appropriately trained database administrator and an understanding of update procedures by range and airspace managers in the field. Additionally, so that they can place appropriate demands upon it, range and airspace managers at field and headquarters levels must become familiar with the contents of the database and the data retrieval interfaces provided with it. Finally, as motivation to keep the system updated, data sources must perceive that the database is used advantageously in addressing critical issues. It is axiomatic that a system perceived to be unused will also be poorly maintained.

THE JOINT MISSION FRAMEWORK

Joint Missions
    Operational Objectives
        Operational Tasks

Deny enemy national leaders the means of conducting military operations and controlling their nations
    Destroy facilities associated with enemy's national and military leadership
        Destroy leadership and security facilities
        Destroy/damage key directing organs and leadership cadres
    Destroy/disable enemy communications networks and control systems
        Disrupt/destroy key communications nodes
        Sever landlines
        Disrupt/disable space-based satellite stations
        Disrupt/disable fixed satellite ground stations
        Disrupt/disable key telephone switching centers
    Destroy/disable war-supporting industries and infrastructure
        Damage/disrupt enemy's war-supporting industry
        Disrupt national petroleum, oil, and lubricants (POL) production, storage, distribution
        Disrupt national transportation system
        Disrupt national power generation and distribution

Deny the enemy the ability to operate aerospace forces and other air defense forces
    Control friendly airspace
        Deconflict friendly traffic
        Identify and track enemy aerial objects

Appendix A—continued

Joint Missions
Operational Objectives
Operational Tasks, continued

Control friendly space
Defend friendly space operations
Establish warning and surveillance systems
Counter enemy ballistic missiles
Destroy transporter-erector-launchers (TELs) in garrisons and assembly areas
Destroy TELs in the field and disrupt operations
Destroy tactical ballistic missile storage areas
Destroy fixed tactical ballistic missile launchers
Defeat attacking ballistic missiles
Warn friendly forces of attack (passive defense)
Destroy ballistic missiles in flight (active defense)
Defeat enemy air attacks
Destroy aircraft in flight
Destroy cruise missiles in flight
Disrupt sensors on enemy aircraft and weapons
Degrade enemy command and control of air forces and integrated air defense
Destroy mobile command posts
Destroy/disrupt airborne command, control, and surveillance platforms
Disrupt communications
Destroy command bunkers and other critical nodes
Destroy/disrupt electronic warfare/ground controlled intercept (EW/GCI) radars
Suppress enemy space-based defenses and offensive capabilities
Destroy/disable space-based space associated facilities
Destroy/disable ground-based space associated facilities

Appendix A—continued

Joint Missions

Operational Objectives

Operational Tasks, continued

Suppress enemy surface-based defenses

Destroy tracking and engagement radars

Destroy mobile surface-to-air missile (SAM)
launchers and anti-aircraft guns

Destroy fixed SAM launchers

Suppress generation of enemy air sorties

Destroy aircraft in the open or in revetments

Destroy key hardened support facilities

Destroy aircraft in hardened shelters

Destroy/damage runways and taxiways

Deny the enemy the ability to operate ground forces

Destroy/demoralize and render ineffective armies in the field

Delay/damage enemy forces and logistics support in
the rear

Disrupt/destroy enemy forces day and night

Degrade enemy command and control of ground
forces

*Evict armies from designated areas, occupy terrain as
necessary*

Deny fire support to enemy defenders

Degrade enemy command and control of ground
forces

Overrun enemy defensive positions

Gain entry into a region

Control enemy forces after surrender

Halt invading armies

Provide fire support to friendly forces in close
contact with enemy ground forces

Delay/destroy/disrupt lead units of invading armies

Delay/damage enemy forces and logistics support in
the rear

Appendix A—continued

Joint Missions
> *Operational Objectives*
> > Operational Tasks, continued

Deny the enemy the ability to operate naval forces and maritime assets
> *Destroy or deny the use of naval support facilities*
> > Destroy ports, logistics facilities, and anchorages
> > Destroy naval command bunkers
> > Disrupt communications and maritime navigation systems
> > Destroy shipborne command posts
>
> *Interdict and control naval combatants and maritime traffic*
> > Disrupt choke points and anchorages
> > Destroy/disable surfaced submarines
> > Degrade/confuse submarine sensors
> > Degrade/confuse shipborne sensors
> > Destroy/disable surface ships at sea or in port

Deny the enemy the capability to produce, store, or deliver weapons of mass destruction (WMD)
> *Destroy facilities producing and storing weapons of mass destruction*
> > Destroy factories and weapon storage sites
> > Deny access to key sites
>
> *Destroy means of delivering weapons of mass destruction*
> > Defeat attacking ballistic missiles
> > Counter enemy ballistic missiles
> > Suppress generation of enemy air sorties
> > Defeat enemy air attacks
>
> *Deter the use of opposing weapons of mass destruction*
> > Ensure survivability of U.S. nuclear weapons and their control
> > Maintain credible threat of retaliation
> > Ensure U.S. ability to operate in WMD environment

Deploy and support forces
> *Deploy forces, support assets, and supplies to theaters of military operations*
> > Provide reconnaissance, surveillance, command and control and attack assessment products

Appendix A—continued

Joint Missions
    Operational Objectives
        Operational Tasks, continued
            Provide communications support
            Provide navigation, geopositioning, and
            weather data
            Conduct at-sea refueling and replenishment
            Sealift personnel and materiel into theaters of
            military operations
            Conduct aerial refueling
            Rescue personnel
            Airlift personnel and materiel into theater
            of operations
    Deploy forces, support assets, and supplies within theaters of
    military operations
            Conduct aerial refueling
            Conduct at-sea refueling and replenishment
            Provide navigation, geopositioning, and
            weather data
            Provide communications support
            Provide reconnaissance, surveillance, command and
            control and attack assessment products
            Rescue personnel
            Airlift personnel and materiel in theater
            of operations
            Sealift personnel and materiel in theaters of
            military operations
            Deploy and redeploy troops within theater
Ensure the implementation of peace agreement/cease-fire
    Ensure disarmament of factions
            Ensure withdrawal/cantonment/destruction of
            heavy weapons
            Seize/destroy illegal weapon caches
            Deny major movements of arms into and
            within territory
    Separate factions
            Observe activities/movements of factions
            Deploy U.S./UN forces in territory between factions

Appendix A—continued

Joint Missions

Operational Objectives

Operational Tasks, continued

Prevent/neutralize attacks of one faction
against another

Support adherence to the agreement

Ensure exchanges of prisoners of war (POWs),
casualties

Support care and repatriation of refugees

Ensure resolutions to implementation disputes at
local level

Support the resolution and punishment of violations

Establish and defend safe areas

Defend safe areas against internal threats

Locate/monitor activities of violent factions

Prevent or eliminate terrorist attacks

Eliminate snipers, particularly in urban terrain

Eliminate SAMs, particularly in urban terrain

Reduce/clear mines/minefields

Protect key facilities/supplies from sabotage

Maintain law and order within safe area

Ensure the enforcement of local laws/regulations

Establish/reconstitute local police authorities

Deter/discourage banditry

Ensure the dispersal, containment, or elimination
of crowds

Protect safe areas against external threats

Destroy/neutralize hostile artillery, mortars

Rescue personnel

Deny infiltration

Disrupt and stop infantry and armor attacks

Disrupt and stop air attacks/establish "no fly" zones

Establish positions at key sites nearby safe areas

Destroy/neutralize key sites

Joint Missions

Operational Objectives

Operational Tasks, continued

Gain control of movement across and within borders

Ensure proper flow of goods and personnel across international borders

Find/monitor key illegal supply and infiltration routes

Disrupt transportation of unauthorized goods and confiscate/destroy

Locate and prevent entry of unauthorized personnel

Maintain freedom of movement on key routes

Protect convoys of supplies/personnel in unsecure areas

Reduce/clear mines and remove roadblocks

Protect critical lines of communication and debarkation points

Gain information superiority

Degrade enemy C3ISR

Penetrate enemy C3ISR systems with cyber attacks

Destroy/disrupt enemy C3ISR assets with physical attack

Gain knowledge of enemy intelligence operations

Degrade enemy picture of battlespace

Protect coalition C3ISR systems

Deny enemy knowledge of friendly intelligence operations

Neutralize enemy C3ISR penetrations

Protect C3ISR assets from physical attack

Establish continuous, fused picture of battlespace

Gain support of local populace

Ensure provision of essential goods and services

Distribute food and water

Establish medical and dental care

Establish temporary shelters

Establish public information/community outreach campaign

Ensure information dissemination

Appendix A—continued

Joint Missions
>   *Operational Objectives*
>>   Operational Tasks, continued

>>   Establish and support community
>>   development programs

Reconstitute civil authority and infrastructure
>   *Ensure reconstitution of government*
>>   Support plebiscites, referenda, and/or elections
>>   Support reconstitution of all branches of government
>>   Support reconstitution of judiciary and penal system
>>   Support establishment of local political bodies
>   *Support government provision of needs of its people*
>>   Promote administration and finance functions
>>   Promote public health, safety, welfare, and
>>   education services
>>   Ensure food supplies and availability of
>>   agriculture components
>>   Promote trade and commerce functions
>   *Support repair of key components of national infrastructure*
>>   Establish essential transportation infrastructure
>>   Establish/support local defense forces

Render humanitarian assistance
>   *Ensure basic services*
>>   Establish medical and dental care
>>   Distribute food and water
>>   Establish temporary shelters
>   *Protect delivery of food and medical supplies to*
>   *distribution points*
>>   Protect ports of entry, storage areas, and key
>>   distribution points
>>   Protect relief ships
>>   Protect relief flights
>>   Protect convoys
>   *Rescue civilians in distress*
>>   Ensure immediate medical attention to the injured
>>   Rescue persons trapped in collapsed structures
>>   Rescue persons in areas of difficult ingress/egress

MISSION/SORTIE DEFINITIONS USED IN THE ANALYSIS

Appendix B

Mission/sortie Definitions Used in the Analysis

Group	Sortie ID	Full Name	Category[a]	Description
Basic flying, non-RAP	INS	Instrument	Basic	Sortie designed to achieve proficiency in instrument flying.
	AHC	Aircraft handling characteristics	Basic	Training for proficiency in utilization and exploitation of the aircraft flight envelope, consistent with operational and safety constraints, including, but not limited to, high/maximum angle of attack maneuvering, energy management, minimum time turns, maximum/optimum acceleration and deceleration techniques, and confidence maneuvers. Sortie may also include basic aircraft navigation and instrument approaches. May be referred to as a Defensive Tactics (DT) sortie in bomber contexts.
	CP	Crew proficiency	Basic	Sortie designed to achieve proficiency in basic flying skills; used in aircraft types with nonaircrew mission crew positions.
	CON	Contact	Basic	Sortie designed to achieve proficiency in helicopter takeoff and landing patterns.
Basic combat	BFM	Basic fighter maneuver	Basic	Sortie designed to apply aircraft handling skills to gain proficiency in recognizing and solving range, closure, aspect, angle off, and turning room problems in relation to another aircraft to attain a position from which weapons may be launched or to defeat weapons employed by an adversary. Scale is 1 v. 1. Maneuvers are within visual range.
	ACM	Air combat maneuver	Basic	Sortie designed to achieve proficiency in element formation maneuvering and the coordinated application of BFM to achieve a simulated kill or effectively defend against one or more aircraft from a pre-planned starting position. For range and airspace requirements, scale is assumed to be 2 v. x. Maneuvers are generally within visual range.
	BSA	Basic surface attack	Basic	Sortie designed to achieve proficiency in medium/low altitude tactical navigation and air-to-surface weapon delivery events.

Appendix B—continued

Group	Sortie ID	Full Name	Category[a]	Description
Basic combat, continued	CSS	Combat skills sortie	Basic	Sortie designed to achieve proficiency in selected events; events are accomplished independently (as building blocks) rather than being integrated in a specific combat scenario.
Fighter, air-to-air	RFUEL	Refueling	Basic	An EC-130 mission dedicated to achieving proficiency in air refueling.
	DCA	Defensive counter air	Applied	Sortie designed to develop proficiency in defensive counter-air tactics. For the purpose of determining range and airspace requirements, the sortie is assumed to be on a 4 v. x scale. Full vertical dimension is used to accomplish tactical objectives.
	OCA	Offensive counter air	Applied	Sortie designed to develop proficiency in air-to-air offensive counter-air tactics. For the purpose of determining range and airspace requirements, the sortie is assumed to be on a v. x scale. Full vertical dimension is used to accomplish tactical objectives.
	OCA ANTI-HELO	Anti-helicopter mission	Variant	A-10 mission to gain proficiency in attacks against opposing helicopters.
Fighter, air-to-ground	SAT FTR	Surface attack tactics (fighter)	Applied	Sortie designed to develop proficiency in surface attacks against a tactical target; should include air or ground threat. Although not specified in AFI 11-2 series publications, the sortie should include tactical navigation events during ingress and egress. For the F-117, includes vertical navigation events.
	SAT FTR GD OPP	SAT (fighter) ground threat opposed	Variant	SAT variant using ground threat emitters to provide simulated opposition.
	SAT FTR AIR OPP	SAT (fighter) air threat opposed	Variant	SAT variant using red air assets to provide simulated opposition.

Appendix B—continued

Group	Sortie ID	Full Name	Category[a]	Description
Fighter, air-to-ground, continued	SAT FTR LIVE	SAT (fighter) with live ordnance	Variant	SAT sortie with delivery of live weapons.
	SAT FTR CLASS OPS	SAT (fighter) with classified operations	Variant	F-117 variant that includes operations to be performed on a classified range.
	CAS	Close air support	Applied	Sortie flown in support of ground forces under the control of a forward air controller, either air or ground.
	SEAD-C	Suppression of enemy air defense, conventional	Applied	Sortie designed to develop proficiency in suppression of enemy air defenses using conventional air-to-ground weapons.
	SEAD	Suppression of enemy air defense	Applied	Sortie designed to develop proficiency in suppression of enemy air defenses using antiradiation weapons.
	FAC-A	Forward air control	Applied	Sortie flown to provide airborne forward air control of armed attack fighters in support of actual or simulated ground fighters.
Bomber	SAT BOMB	Surface attack tactics (bomber)	Applied	Sortie designed to develop proficiency in surface attacks against a tactical target; should include air or ground threat. Sortie should include tactical navigation events during ingress and egress.
	SAT BOMB INRT LO	SAT (bomber) low altitude	Variant	SAT BOMB variant in which actual release of inert ordnance occurs from low altitude.
	SAT BOMB INRT HI/MED	SAT (bomber) hi/medium altitude	Variant	SAT BOMB variant in which actual release of inert ordnance occurs from medium or high altitude.

Appendix B—continued

Group	Sortie ID	Full Name	Category[a]	Description
Bomber, continued	SAT BOMB LIVE	SAT (bomber) with live ordnance	Variant	SAT sortie with delivery of live weapons.
	SAT BOMB SIM	SAT (bomber) simulated release	Variant	SAT BOMB variant in which release of weapons is simulated.
	SAT BOMB MARI-TIME	SAT (bomber) maritime	Variant	SAT BOMB variant flown to practice mining a maritime target.
Other	MSN HC130	Mission, HC-130	Applied	Combat scenario profile that relates to the requirements of the unit's DOC statement.
	MSN HC130 WATER	Mission, HC-130 over water	Variant	HC-130 search and rescue mission variant performed over water.
	MSN E8C	Mission, JSTARS	Applied	Combat scenario profile that relates to the requirements of the unit's DOC statement.
	MSN E8C TM	JSTARS mission with terrain masking	Variant	JSTARS mission sortie that includes terrain masking.
	MSN E8C RETRO	JSTARS mission with retrograde event	Variant	JSTARS mission sortie that includes a combat separation event (departure from orbit and rapid descent to gain airspeed).
	MSN E3	Mission, AWACS	Applied	Combat scenario profile that relates to the requirements of the unit's DOC statement.
	MSN EC130H	Mission, Compass Call	Applied	Combat scenario profile that relates to the requirements of the unit's DOC statement.

Appendix B—continued

Group	Sortie ID	Full Name	Category[a]	Description
Other, continued	MSN EC130E	Mission, ABCCC	Applied	Combat scenario profile that relates to the requirements of the unit's DOC statement.
	MSN RC135	Mission, Rivet Joint	Applied	Combat scenario profile that relates to the requirements of the unit's DOC statement.
	MSN U2	Mission, U-2	Applied	Combat scenario profile that relates to the requirements of the unit's DOC statement.
	MSN HH60G	Mission, HH-60G	Applied	Combat scenario profile that relates to the requirements of the unit's DOC statement.
	MSN UAV	Mission, UAV	Applied	Combat scenario profile that relates to the requirements of the unit's DOC statement.
Combined	AWACS A-A	AWACS with air-to-air	Combined	Sortie designed to exercise AWACS with air-to-air fighters.
	AWACS EC	AWACS electronic combat	Combined	Sortie designed to exercise AWACS employment in conjunction with a SEAD mission.
	CSAR	Combat search and rescue	Combined	Small multi-MDS exercise (SMME) that combines HH-60 MSN, SAT (F-16, A-10, F-15E), HH-130 MSN, and/or UAV MSN roles.
	FP/ SWEEP	Force protection/ sweep exercise	Combined	SMME that combines OCA (F-15C, F-16, F-15E, F-22) and SAT (F-15E, F-16, F-117, A-10, B-1, B-2, B-52) roles. Targets may be on or off range.
	LFE	Large force engagement	Combined	Flag or other large training exercise involving multiple flights of aircraft types in a variety of roles; simulates the scale and complexity of actual combat. Combines SAT, OCA, AWACS MSN, JSTARS MSN, UAV MSN, ABCCC MSN, CAS, CSAR, and/or Rivet Joint MSN roles.

[a] *Basic* sorties are building-block exercises that are used to train fundamental flying and operational skills. *Applied* sorties are intended to more realistically simulate combat operations, incorporating intelligence scenarios and threat reaction events. *Variants* are subdivisions of Ready Aircrew Program (RAP) sorties, constructed for this study, that differ significantly from each other in their infrastructure requirements. *Combined* sorties are structured to bring together several MDS, performing different operational roles, in a single training mission.

RANGE AND AIRSPACE CHARACTERISTICS

The following table lists the range and airspace characteristics captured for our analysis. As indicated, some information was captured only for MDS/sortie types, some only for available assets (existing ranges and airspace), and some for both requirements and available assets. Some characteristics are in text form (e.g., scheduling agency), some are in numerical form (e.g., length in nautical miles), and some are in binary (yes/no) form (e.g., authorization to dispense chaff). Binary characteristics are punctuated using a question mark in the characteristics column. Binary characteristics are interpreted as indicating whether the item is required (in a requirements array) or authorized/available (in an available assets array).

Threats are listed as a separate infrastructure type. However, threat emitters and communications jammers must be installed on a route, area, or range. In the database, threat requirements appear only once in any given MDS/sortie-type requirements array. However, threat infrastructure availability is recorded for each route, maneuver area, and range.

Appendix C

Range and Airspace Characteristics

Infrastructure Type	Characteristics	Require-ments	Available Assets
Low-level routes	Name/designation		X
	Reporting agency		X
	Scheduling agency		X
	Point of contact for scheduling agency		X
	Commercial phone for POC[a]		X
	DSN phone for POC		X
	Entry latitude (decimal degrees)		X
	Entry longitude (decimal degrees)		X
	Exit latitude (decimal degrees)		X
	Exit longitude (decimal degrees)		X
	Alternate entry points?		X
	Alternate exit points?		X
	Open 24 hours?		X
	Charted opening time		X
	Charted closing time		X
	Days per week		X
	Percentage of operating hours unavailable due to maintenance		X
	Percentage of operating hours used by non- ACC users		X
	Percentage of operating hours used by ACC users		X
	Flight spacing (minutes)		X
	Length (nm)		X
	Width (nm)		X
	Floor (ft above ground level [AGL])		X
	Ceiling (ft AGL)		X
	Route time (minutes)	X	
	Speed (knots)	X	
	Minimum width (nm)	X	
	Minimum length (nm)	X	
	Maximum floor	X	
	Minimum ceiling	X	
	Terrain-following operations?	X	X
	Segment below 300 ft?	X	X

Appendix C—continued

Infrastructure Type	Characteristics	Require- ments	Available Assets
Low-level routes, continued	25-nm segment cleared up to 5000 ft?	X	X
	Instrument meteorological conditions (IMC)-capable:	X	X
	Percentage of route required to be mountainous	X	X
	Training route leads into/passes thru MOA or warning area?	X	X
	Name/designation of adjoining MOA or warning area		X
Maneuver areas	Name/designation		X
	Reporting agency		X
	Scheduling agency		X
	POC for scheduling agency		X
	Commercial phone for POC		X
	DSN phone for POC		X
	Latitude at center (decimal degrees)		X
	Longitude at center (decimal degrees)		X
	Open 24 hours?		X
	Charted opening time		X
	Charted closing time		X
	Days per week		X
	Percentage of operating hours unavailable due to maintenance		X
	Percentage of operating hours used by non-ACC users		X
	Percentage of operating hours used by ACC users		X
	Width		X
	Length		X
	Floor (ft)		X
	Floor type (AGL or mean sea level [MSL])		X
	Ceiling (ft MSL)		X
	Minimum width (nm)	X	
	Minimum length (nm)	X	
	Maximum floor (ft AGL)	X	
	Minimum ceiling (ft MSL)	X	

Appendix C—continued

Infrastructure Type	Characteristics	Require-ments	Available Assets
Maneuver areas, continued	Lowest floor for an altitude block (ft MSL)	X	
	Highest ceiling for an altitude block (ft MSL)	X	
	Minimum altitude block required (ft)	X	
	Chaff?	X	X
	Flares?	X	X
	Over land?	X	X
	Over water?	X	X
	Over mountains?	X	X
	Air-air communications?	X	X
	Air-ground communications?	X	X
	Datalink?	X	X
	Adjoining orbit?	X	X
	Name/designation of adjoining orbit		X
	Access to air-ground range?	X	X
	Name/designation of adjoining range		X
	ACMI?	X	X
	Supersonic operations?	X	X
Ranges	Name/designation		X
	Alternate name		X
	Complex		X
	Reporting agency		X
	Scheduling agency		X
	Scheduling base		X
	POC for scheduling agency		X
	Commercial phone for POC		X
	DSN phone for POC		X
	Latitude at center (decimal degrees)		X
	Longitude at center (decimal degrees)		X
	Open 24 hours?		X
	Charted opening time		X
	Charted closing time		X
	Days per week		X
	Percentage of operating hours unavailable due to maintenance		X

Appendix C—continued

Infrastructure Type	Characteristics	Require- ments	Available Assets
Ranges, continued	Percentage of operating hours used by non-ACC users		X
	Percentage of operating hours used by ACC users		X
	Width of restricted airspace (nm)		X
	Length of restricted airspace (nm)		X
	Ceiling of restricted airspace (ft MSL)		X
	Width of weapon safety footprint area		X
	Length of weapon safety footprint area		X
	Restricted airspace minimum width (nm)	X	
	Restricted airspace minimum length (nm)	X	
	Restricted airspace minimum ceiling (ft MSL)	X	
	Weapon safety footprint area minimum width (nm)	X	
	Weapon safety footprint area minimum length (nm)	X	
	Conventional circles?	X	X
	Strafe pits?	X	X
	Strafe targets 30mm authorized?	X	X
	Number of bomb targets scored simultaneously	X	X
	Lighted targets?	X	X
	Vertical targets?	X	X
	Tactical target array?	X	X
	Second tactical target array separated by 30nm from the first array?	X	X
	Urban target array?	X	X
	Ordnance type (inert, live, or both)	X	X
	Number of laser targets required	X	X
	Number of infrared-significant (heated) targets	X	X
	Number of radar-significant targets	X	X
	Scoring no drop?	X	X
	Laser spot scoring?	X	X
	Night scoring?	X	X
	Scoring with 1-meter accuracy?	X	X

Appendix C—continued

Infrastructure Type	Characteristics	Requirements	Available Assets
Ranges, continued	Scoring available within 15 seconds of impact?	X	X
	Chaff/flare/ECM pods?	X	X
	Illumination flares?	X	X
	Attack heading variable by 90 degrees?	X	X
	Secured to allow classified operations?	X	X
	Night vision goggles?	X	X
	Part of range over water?	X	X
	Range control officer (RCO)?	X	X
Threats	Number of required threat emitters	X	X
	Multiple threat emitter?	X	X
	FSU area defense emitter?	X	X
	Non-FSU threat emitter?	X	X
	Transportable threat emitter?	X	X
	Post-mission threat reaction debrief capability?	X	X
	Reactive threat emitter system?	X	X
	Smokey SAMs?	X	X
	Radar jammer?	X	X
	Communications jammer?	X	X
Orbits	Name/designation		X
	Type (refueling, mission)		X
	Reporting agency		X
	Scheduling agency		X
	POC for scheduling agency		X
	Commercial phone for POC		X
	DSN phone for POC		X
	Entry latitude (decimal degrees)		X
	Entry longitude (decimal degrees)		X
	Exit latitude (decimal degrees)		X
	Exit longitude (decimal degrees)		X
	Open 24 hours?		X
	Charted opening time		X
	Charted closing time		X

Appendix C—continued

Infrastructure Type	Characteristics	Require-ments	Available Assets
Orbits, continued	Days per week		X
	Percentage of operating hours unavailable due to maintenance		X
	Percentage of operating hours used by non-ACC users		X
	Percentage of operating hours used by ACC users		X
	Length (nm)		X
	Width (nm)		X
	Floor (ft MSL)		X
	Ceiling (ft AGL)		X
	Minimum width for refueling (nm)	X	
	Minimum length for refueling	X	
	Maximum floor for refueling (ft MSL)	X	
	Minimum ceiling for refueling (ft)	X	
	Minimum floor for refueling altitude block (ft)	X	
	Maximum ceiling for refueling altitude block (ft)	X	
	Altitude block required for refueling (ft)	X	
	Minimum width for mission (nm)	X	
	Minimum length for mission (nm)	X	
	Maximum floor for mission (ft MSL)	X	
	Minimum ceiling for mission (ft)	X	
	Altitude block required for mission (ft)	X	
	Percentage of the orbit/track over mountainous terrain	X	X
	Radiatable air-to-ground or artillery range at 90–150 nm from orbit?	X	X
	Direct access to Army maneuver area or air-to-ground range?	X	X
	Air-to-air range 60–120 nm from orbit?	X	X
	Dedicated air-to-air frequency?	X	X
	Dedicated air-ground frequency?	X	X
	ABCCC training capsule?	X	X
	JTIDS datalink needed?	X	X
	Surveillance control datalink (SCDL)?	X	X

Appendix C—continued

Infrastructure Type	Characteristics	Require-ments	Available Assets
Orbits, continued	Communications system operator training (CSOT) capability?	X	X
	JSTARS workstation?	X	X
Other	Threat air-to-air fighter?	X	
	Any air-to-air fighter?	X	
	Any air-to-ground fighter?	X	
	Threat air-to-ground fighters?	X	
	Heavy bomber?	X	
	Tanker?	X	
	E-3 (AWACS)?	X	
	E-8C (JSTARS)?	X	
	EC-130H (ABCCC)?	X	
	Ground FAC?	X	
	Ground control intercept?	X	
	Ground movers?	X	
	Post-mission truth data?	X	
	Landing zone?	X	

aPoint of contact.

SIZING THE AIRSPACE FOR AIR-TO-GROUND AND AIR-TO-AIR TRAINING SORTIES

The sizing of maneuver area requirements for air-to-ground and air-to-air training sorties is based on the most demanding normal training scenarios described to us by Air Force pilots.[1] From their descriptions, we identified a canonical sequence of maneuvers and related speeds, turning gravitational forces, bank angles, etc. We then built simulations using these inputs to compute flight trajectories dynamically. Scenario parameters were stored in an MS Excel workbook, allowing easy modification and recomputation. In addition to computing nominal trajectories, the simulations allow leeway for effects such as underbanked turns (resulting in larger turn radii that use additional space).

These simulations were developed during and following a round of visits to ACC field units in which we obtained estimates of area requirements from experienced aircrews. The simulation results generally confirm the estimates we received. Ideally, we would have completed a second round of visits to field units to validate and fine-tune the area requirements predicted by the simulations. However, limited resourcing in the project precluded this step. Because we were unable to fully validate the simulation results, maneuver area requirements reflected in the range and airspace database are based primarily on aircrew judgments rather than the simulations.

---

[1]An important defect of this approach is the small sample of pilots used to obtain the input. Engaging a larger number of pilot reviewers required too much time on their part to be feasible.

However, we document the simulations in this appendix so that they can be used in any future efforts to refine infrastructure requirements or to define requirements for new systems.

BSA MANEUVER AREA FOR THE F-15E AND F-16

The BSA maneuver area requirement for the F-15E and F-16 is built up from the flight-path pattern shown in Figure D.1. From the indicated starting point, a straight run-in is made to a release point. Then a turn-away is made, followed by a run back to a repositioning maneuver. The figure is actually a superposition of the two most stressing (on the area size) deliveries that are made. Distance required in front of the target is determined by release of the longest-range weapon the system carries. Distance to the rear of the target is

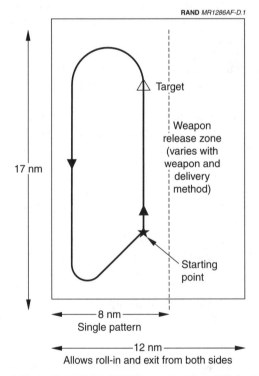

Figure D.1—F-15/F-16 BSA Maneuver Area Flight Path

determined by a release essentially over the target, so the turn-away commences at the target. Components that need to be specified are listed in Table D.1. The relationships of these parameters to the flight trajectory are shown in Figure D.2.

Table D.1

BSA Maneuver Area Parameters

Symbol	Definition
B_1	Buffer allowed for turnout at top of pattern
B_2	Buffer (from center of target area) to edge of maneuver area
B_3	Buffer from edge of maneuver area during runback for additional pass
B_4	Buffer allowed for turn back to reposition bottom of pattern
R_T	Radius of tactical turn after weapon release
R_2	Radius of turns on/off the mapping leg
D_{RP}	Distance from center of target array to most distant release point
D_{RI}	Distance to run in before release
D_{TB}	Distance beyond target before turn back to map for additional pass
D_E	Distance beyond target, after release, before tactical turn
W_{TGT}	Width (length) of target array
D_{CK}	Distance required to perform system checks during return leg
D_{MAP}	Mapping distance
α_{MAP}	Mapping angle

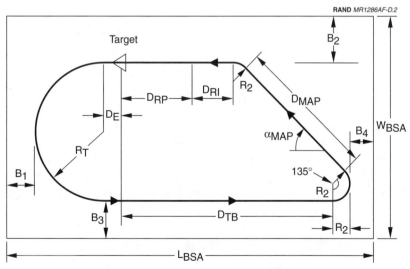

Figure D.2—F-15E/F-16 BSA Maneuver Area Flight Path with Parameters

When the maneuver is used on a tactical target array, the area requirement must consider the possibility that the target may be in any corner of the array. This serves to add W_{TGT} to both the length and width. The total length of the area is then given by

$$L_{BSA} = B_1 + R_T + D_E + D_{TB} + R_2 + B_4 + W_{TGT} \quad .$$

The total width of the area is given by

$$W_{BSA} = B_2 + 2 \times R_T + B_3 + W_{TGT} .$$

If turnouts to either the right or the left are to be possible, the width is instead given by

$$W_{BSA} = 2 \times (B_3 + 2 \times R_T) + W_{TGT} .$$

Computation of the Parameters

R_T, the tactical turn radius, is computed based on either a speed and bank angle or a speed and gravitational force.

D_E is determined by safe escape (from bomb fragments) considerations.

D_{RP} is directly specified, based on the longest release range across munitions used by the MDS.

D_{RI} is associated with a minimum run-in time at tactical speed.

R_2 is based on turning at a specified number of G's at a speed somewhat below the tactical speed.

The formula for the length of the BSA area includes D_{TB}, which must be calculated. This can be done using two equations that express the fact that an aircraft flying the entire loop ends up back where it starts. The equations give the total lengthwise and widthwise displacements around the loop:

$$0 = D_{TB} + R_2 \times \sin(135°) - D_{MAP} \times \cos(45°)$$
$$- R_2 \times \sin(45°) - D_{RI} - D_{RP}$$

$$0 = -2 \times R_T + R_2 \times [1 - \cos(135°)] + D_{MAP} \times \sin(45°)$$
$$+ R_2 \times [1 - \cos(45°)] \; .$$

These may be solved to give

$$D_{TB} = 2 \times (R_T - R_2) + D_{RI} + D_{RP}$$

$$D_{MAP} = 2\sqrt{2} \times \left(R_T - R_2\right) \; .$$

Adjustments may be needed. First, the return leg D_{TB} is used not only for repositioning but also to make certain system checks, and this requires a minimum time (or equivalent distance). If the distance to perform system checks, D_{CK}, is greater than this, then D_{TB} in the figure needs to be replaced by D_{CK}.

A second adjustment is needed if D_{MAP} turns out to be too short for a minimum mapping time. It may be acceptable to overshoot the target attack axis, to prolong the mapping. However, when this is unacceptable, mapping can be done at an angle less than 45° with respect to the major axis, resulting in a longer D_{TB}. When D_{MAP} is specified as the minimum acceptable mapping distance (and is greater than the value

$$2\sqrt{2}\left(R_T - R_2\right)$$

associated with $\alpha_{MAP} = 45°$), the loop displacement equations can be solved for αMAP and D_{TB}:

$$\sin(\alpha_{MAP}) = 2 \times (R_T - R_2) \, / \, D_{MAP}$$

$$D_{TB} = \text{SQRT}\left[D_{MAP2} - 4\left(R_T - R_2\right)^2 \right] + D_{RI} + D_{RP}.$$

However, in typical cases mapping time will not be a problem. For example, suppose R_T is based on a 45 bank at 480 kts, and R_2 is based on a 4-G turn at 420 kts. Then D_{MAP} for the nominal 45 run comes to 7.6 nm, affording more than a full minute of mapping time at 420 kts. This is greatly in excess of any requirement.

Buffers

Additional buffers, outside of the nominally required maneuver space, must be included for safety reasons. One contributor to a buffer requirement is the tendency to underbank the tactical turn, resulting in an increase to the radius R_T. In general, going from a level turn at bank angle θ to one at a smaller angle θ' will increase R_T by an amount

$$R_T \times (\tan \theta - \tan \theta') / \tan \theta .$$

The increased width, from a 180 turn, is twice this value. Reducing bank angle from 45° to 40° at 480 kts increases the width requirement by 1.29 nm. On the final turn, a 45 turn at 4 Gs and 420 kts, a 10° underbank (from 75.5° to 65.5°), would result in an increased turn radius of 0.51 nm. If the turn was not started early (probably unlikely), this would result in an overshoot of 0.15 nm, since only 45° of turn is involved.

In general, an additional buffer of 1 nm to each side of the area is needed, if only so that pilot concentration is not degraded by the need to fly extremely close to area boundaries.

Table D.2 shows a set of realistic input values and calculated outputs for BSA area sizes.[2]

BSA MANEUVER AREA FOR THE A-10

The A-10 does not do a radar map, so its maneuver area is more rectangular. There are several alternative trajectories for BSA, and the one that has the largest infrastructure requirement entails a safe escape maneuver running straight at the target until 3 seconds after weapon impact. This is shown in the diagram in Figure D.3. In this figure, the target is in the lower right. The actual trajectory is not circular on the left; it consists of two quarter-circles of differing radii and a short straight segment for acceleration. Computation of the dimension requirements is as indicated in Table D.3.

---

[2]MS Excel spreadsheets performing the calculations for this and other worksheet tables in this appendix are available from the authors.

Table D.2

F15-E/F-16 BSA Worksheet

Input/ Output	Variable	Short Name	Description	F-16 Value	F-15E Value
Input	TacSpd		Tactical speed (kts)	500	480
	TacTrnBnk		Tactical turn bank angle (deg)	60	45
	TacTrnUBnk		Tactical turn underbank angle, for leeway calculation (deg)	10	5
	MapSpd		Speed used on map leg (kts)	420	420
	MapTurnG		Gs for turns to/from map	4	4
	MapTurnUBnk		Underbank angle for map turn (deg)	10	10
	TAfterDet		Time between deton- ation and start of tactical turn	3	3
	DRelease		Maximum munition release distance (nm)	5	5
	TRunIn		Desired run-in time before munition release	30	30
	TMapRqd		Minimum mapping time required	30	30
	Buffer	B_1-B_4	Additional buffer (nm)	1	1
	TgtArraySize		Size of target array (nm)	1	1
Output	TacRad	R_T	Radius of tactical turn	2.10	3.36
	TacUB_leeway		Additional lateral dimen- sion required to account for underbank	1.91	1.29
	Rad2	R_2	Radius of turns to/from map leg	0.66	0.66
	Rad2_lw_len		Leeway for underbank during the turn to the map leg	0.51	0.51
	Rad2_leeway		Leeway for underbank during the 45 deg turn onto target	0.15	0.15
	D_E	D_E	Distance after target before beginning tactical turn	0.42	0.40
	DRunIn	D_{RI}	Run-in distance before release, after turn to target	2.47	2.37
	DTurnAround	D_{TB}	nm	10.35	12.76

Table D.2 —continued

Input/ Output	Variable	Short Name	Description	F-16 Value	F-15E Value
	DMap	D_{MAP}	nm	4.07	7.62
	TMap		sec	35	65
	MapTimeOkay		If FALSE, calculation is invalid	TRUE	TRUE
	AreaLength	L_{BSA}	nm	18.0	21.3
	AreaWidth	W_{BSA}	nm	8.3	10.2

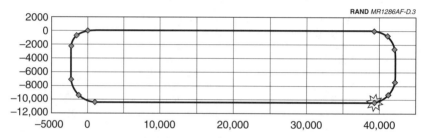

RAND MR1286AF-D.3

Figure D.3—A-10 BSA Maneuver Area Flight Path (feet from target)

Table D.3

A-10 BSA Worksheet

Input/ Output	Variable	Units	Description	Value
Input	A10SpdTurnOff	kts	Speed for turn off target	250
	A10GTurnOff		Gs for turn off target	2
	A10GTurnIn		Gs for turn to crossing leg and turn into target	2.5
	A10CrossSpd	kts	Speed for crossing leg	300
	A10SpdRunIn	kts	Speed during run-in	320
	A10DRunIn	nm	Distance to run-in, after turn in	1
	A10DLaunch	nm	Launch range	5.3
	A10AccelG		Acceleration going from SpdTurnOff to CrossSpd	0.25
	A10TgtArraySize	nm		1
	A10BufferSize	nm		2
Output	ySize	nm	2-sided	9.2
	xSize	nm		14.0

SAT MANEUVER AREA FOR THE F-15E AND F-16

Figure D.4 shows the maneuver area required for SAT. There is an initial point at a distance D_{IP} from the target. The maneuver shown entails a map phase where an angle of θ_{MAP} with respect to the initial line-of-sight (LOS) to the target is maintained while a target picture is built with the side-angle radar. A distance D_{MAP} is flown, based on a desired illumination time. Then a run-in is made toward the target.[3] The aircraft pops up to release at a distance R_L from the target and flies at an angle of θ_{ILLUM} with respect to the target until weapon impact. Then, the attacker might either turn away immediately or extend behind the target before turning back for another run. Figure D.4 shows only the turn-away option. Also, additional attackers may be present. The figure shows only the flight path for the attacker who will go widest, which is the determiner of the required area width.

Table D.4 shows the SAT dimensions implied for these maneuvers, including a buffer zone and allowance for target array size.

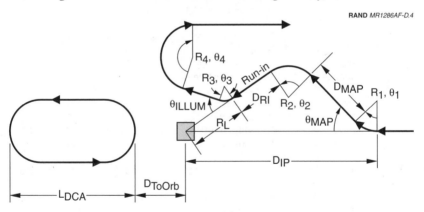

RAND *MR1286AF-D.4*

Figure D.4—F-15E/F-16 SAT Maneuver Area Flight Path

---

[3]In actuality, running straight at the target is not desirable because it adds predictability. It is not necessary, thanks to the mapping. A more realistic tactic can be accounted for in the diagram by increasing D_{MAP} to include a low-level run after mapping is completed. The amount of the increase should be the largest reasonable extension distance that would be used in training.

Table D.4

SAT Loft Attack Worksheet

Input/ Output	Variable	Description	F-15E Value	F-16 Value
Input	distIP	Distance from target to IP[a] (nm)	12	12
	sMap	Speed for mapping, until run-in (kts)	420	420
	theta Map	Angle with respect to target for mapping (deg)	30	30
	gee1	Gs for turn to map	4	4
	ubank1	Underbank angle for gentler turn (deg)	10	10
	tMap	Time on mapping leg (sec)	30	30
	gee2	Gs for turn to run-in	4	4
	ubank2	Underbank angle for gentler turn (deg)	10	10
	sRunIn	Speed for run-in and remainder of engagement	480	500
	rLaunch	Launch range (nm)	5	5
	wpnTOF	Weapon TOF[b]	55	55
	theta Illum	Angle with respect to target for illumination (deg)	60	60
	gee3	Gs for turn to illuminate	4	4
	ubank3	Underbank angle for gentler turn (deg)	10	10
	gee4	Gs for turn to egress	1.41	2.00
	ubank4	Underbank angle for gentler turn (deg)	5	10
	tgtSize	Size of target array (nm)	1	1
	buffer	Buffer required on each side (nm)	2	2
Output	ymax	1/2-width with no underbank (nm)	12.7	10.8
	ymax_ub	1/2-width with underbank included (nm)	14.0	12.6
	width Req	Full width including buffers and target size (nm)	33.0	30.3
	length Req	For single-attack axis, not including DCA orbit	20.2	20.3
	lenBeforeTgt	When including DCA orbit, length = lenBeforeTgtt + distance from target to back of DCA orbit	14.5	14.5
	lenBeyondTgt	When *not* including DCA orbit, length = lenBeforeTgt + lenBeyondTgt	5.7	5.8

[a]Impact point.

[b]Time of flight.

The width requirement is based on the ability to perform this attack to either the right or the left, including allowance for buffers and target array size. The length requirement is that required for a single axis of attack and does not include an allowance for a DCA orbit. When such an orbit is considered, the total length dimension is given by the sum of the distance needed from the IP point to the (lenBeforeTgt) plus the distance beyond the target to the back of the DCA orbit. The plot in Figure D.5 shows the computed flight path for nominal and underbanked turns. The axes depict a Cartesian coordinate system originating at the target location.

The other trajectory that must be considered is for extension beyond the target. This is depicted in Figure D.6.

Table D.5 shows the SAT dimensions implied for these maneuvers, including a buffer zone and allowance for target array size. Alternative value sets may be used to compute trajectories for other MDS.

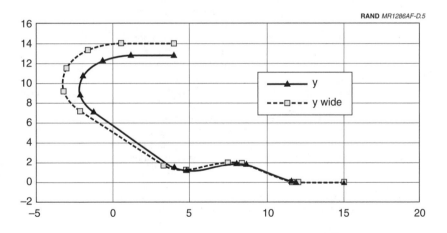

Figure D.5—Computed Flight Paths for SAT Loft Attack (nm from target)

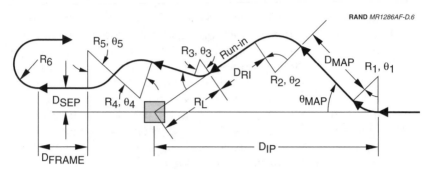

Figure D.6—SAT Extension-Beyond-Target Flight Path

Table D.5

SAT Extension-Beyond-Target Worksheet

Input/ Output	Variable	Units	Description	F-15E Value	F-16 Value
Input	dIP	nm	Distance to IP point (where mapping turn begins)	12	12
	sMap	kt	Speed during map phase (and turns to/from map)	420	420
	gMap	–	Gs for turn to map	1.41	2.00
	thetaMap_SE	deg	Angle off target for map turn	45	45
	tMap	sec	Mapping time	30	30
	gTurnIn	–	Gs for turn to run-in	4	4
	sRunIn	kts	Speed during run-in phase, and remainder of engagement	480	500
	rLaunch	nm	weapon launch range	5	5
	gIllum	–	Gs for turn to "illuminate"	4	4
	thetaIllum_SE	deg	Angle off target for "illumination"	60	60
	wpnTOF	sec	Weapon TOF	55	55
	weave Angle	deg	Angle with respect to range major axis, for weave	45	45

Table D.5 —continued

Input/ Output	Variable	Units	Description	F-15E Value	F-16 Value
	exitFormation Spacing	feet	Tactical spread for formation during radar check	6000	6000
	gWeave	–	Gs to use for turns to/from weave	4	4
	tFrame	sec	radar frame time	24	24
	gHome	–	Gs to use for turn home after radar check	2	2
	tgtArraySize	nm	Size of target array (nm)	1	1
	bufferSize	nm	Buffer required on each side (nm)	2	2
Output	width	nm	Full width including buffers and target size (nm)	14.8	15.0
	length	nm	For single-attack axis, not including DCA orbit	25.7	26.1
	lenBeforeTgt	nm	When including DCA orbit, length = lenBeforeTgt + distance from target to back of DCA orbit	14.5	14.5
	lenBeyondTgt	nm	When *not* including DCA orbit, length = lenBeforeTgt + lenBeyondTgt	11.2	11.6

PUTTING THE PARTS TOGETHER

1. *Evaluate loft attack and extension-beyond-target trajectories.* The loft attack and extension trajectories each imply major and minor axis requirements for the maneuver area. The larger, for each axis, determines the required overall area size for ground-opposed SAT. For air-opposed SAT, the larger of the widths is still relevant. For length, however, the required distances before and after the target should be tracked separately, because a DCA orbit, located beyond the target, will generally be greater than the space required for the attacking aircraft's extension beyond the target.

2. *Provide an area for opposing DCA to orbit.* This adds a square of size L_{DCA} by W_{DCA} to the major axis above. The width W_{DCA} is

determined by an orbit width and buffer, and L_{DCA} is offset from the target by a distance D_{ToOrb}, depending on how the orbit is positioned with respect to the target.

3. *Rotate the area by 90° to generate an area that can be used for multiple axes of attack.* In effect, the desired area is a square with sides equal to the larger of the original area major and minor axes.

SAT MANEUVER AREA FOR THE A-10

A-10 SAT are more fluid and less amenable to a stylized representation than are F-15E and F-16 SAT. A common characteristic of A-10 SAT, however, is that of attacking from a wheel, or circle, that may be centered on the target, or may have the target lying on the circle. This latter case should imply a larger area, assuming the same circle radii.

The area implied by Figure D.7 is a square four times the radius of an attack circle, plus the target array size. However, other considerations, such as approach and egress, may require extending the area in at least one direction. These circles will not be very large. For instance, the radius of a circle based on a 45° banked turn at 250 kts is only 0.64 nm. For a 1-nm square target array size, this computes to an area 3.5 nm on a side.

Realistic A-10 SAT training will additionally include a cross-country leg associated with a call to attack an unplanned target. The area size associated with this requirement will dominate that associated with maneuvers in the immediate target vicinity.

BFM MANUEVER AREA

Figure D.8 shows a notional BFM training sortie. Because BFM sorties can evolve in numerous ways, it makes little sense to diagram any particular set of maneuvers. However, one feature of the figure is important to note—the reset for a second engagement does not return the flight to the original ("fight on") starting point. Having sufficient maneuver area to permit such resets is important to conserve fuel and enable a larger number of engagements to be undertaken.

RAND *MR1286AF-D.7*

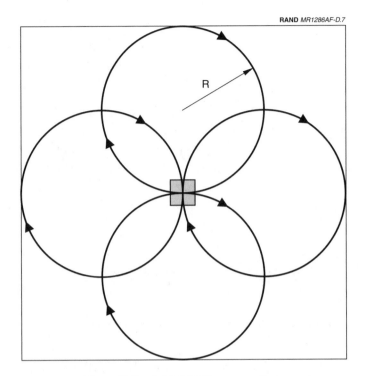

Figure D.7—A-10 SAT Maneuver Area

The general unpredictability of BFM engagements requires that experienced pilot judgment be used to determine the required area size, taking into account both reasonable space for a single engagement and the space required to permit subsequent engagements to begin offset from the middle of the maneuver area.

DCA MANEUVER AREA

In some ways, the DCA area requirement is easier to compute than that needed for BFM, despite the additional complexity and unpredictability associated with DCA. The length requirement is developed from considerations of steps that are for the most part fairly predictable as to the time/distance required for their development. It is not dominated by the more unpredictable course of the

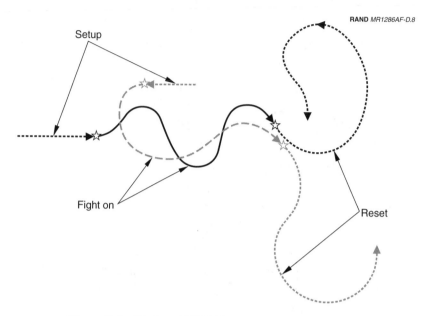

Figure D.8—Notional BFM Maneuver Area Flight Paths

engagement after initial weapon release. The width requirement for
a single engagement is not so predictable. The approach taken here
is to consider the width requirement for maneuvers that develop
with reasonable frequency and which tend to involve greater
amounts of lateral movement. An area with a width that accommo-
dates these maneuvers should be satisfactory for the vast majority of
realistic sorties.

The parameters used in considering DCA area length and width are
shown in Figure D.9 and Table D.6.

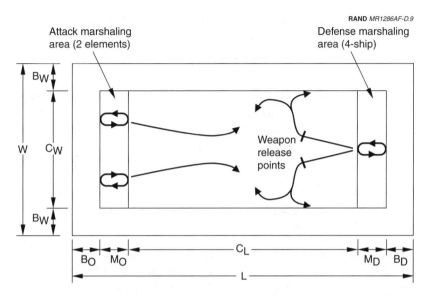

Figure D.9—DCA Maneuver Area Flight Paths

Table D.6

DCA Maneuver Area Parameters

Symbol	Definition
MO	Marshaling area depth for offensive side
MD	Marshaling area depth for defensive side
CW	Combat area width (derived)
CL	Combat area length (derived)
BO	Buffer behind offensive marshaling zone
BD	Buffer behind defensive marshaling zone
BW	Buffer to avoid sideways spillout from the combat zone
L	$B_O + M_O + C_L + M_D + B_D$ (derived)
W	$C_W + 2 \times B_W$ (derived)

Combat Area Length

The combat area length, C_L, is determined by a number of factors. As defensive aircraft come off a cap orbit they must use their radar to build a picture of the enemy forces. This takes a certain time T_B that can be converted, using an average speed during this phase of S_B, to a build distance

$$D_B = T_B \times S_B \ .$$

Next, there is a sorting phase where targeting assignments are made. This takes a time T_S at an average speed of S_S (presumably higher than S_B, assuming that the flight has been accelerating after coming off the cap orbit). It gives a sort distance

$$D_S = T_S \times S_S \ .$$

Sorting should be completed outside of maximum air-to-air missile firing range R_M.

The distance traveled by the offensive force needs to be accounted for as well. It has come off of its marshaling orbits and is traveling at an average speed of S_O. Assume that a time T_R (R for react) passes before the defense leaves the cap orbit. Then the total distance traveled by the offensive force, prior to defensive missile firing, is

$$D_T = S_O \ (T_R + T_B + T_S) \ .$$

This gives a value of C_L given by

$$C_L = D_T + D_B + D_S + R_M \ .$$

Obviously, the parameters that determine C_L will depend on attributes of both the offensive and defensive aircraft involved in the engagement. Also, several variations are possible. For example, the first shot might be taken by the offensive side—the offensive side may have a longer-range air-to-air missile or a doctrine that dictates firing early, before the shot has a high probability of kill. In this case, the distance traveled by the defensive side, prior to first missile firing, would be reduced.

Another variation might require some maneuver by the defenders before firing. The delay associated with this maneuver could be incorporated into the sorting time T_S, so long as the speed S_S is the component of velocity along the primary (length) axis and not the actual speed of the defending aircraft. This is not critical. The critical point is that the actions that must take place prior to initial weapon exchange consume distance and need to be included in the length of the air-to-air template.

Combat Area Width

The width of the combat area should permit realistic tactical behavior. There can be many reasons for moving laterally, including giving the adversary a detection/sorting problem and minimizing exposure to enemy missiles. This discussion explores the enemy missile avoidance issue.

Exposure to enemy missiles is minimized by maximizing the range to the enemy. After firing one's own missile (AMRAAM—advanced medium-range air-to-air missile) one would like to immediately turn away from the adversary to maximize separation, but this is not possible because the adversary must be illuminated with radar until the missile's own radar can acquire the target. The best that can be done is to veer away from the target at an angle that keeps the target near the radar gimbal limit. If this angle is θ and the time required for illumination is T_I then the lateral distance moved by the attacker is approximately given by

$$D_I = S_A \times T_I \times \sin \theta \ ,$$

where S_A is the attacker speed during this phase. This formula is approximate for several reasons. For example, as the target aircraft approaches, the attacker must gradually turn back toward the attacker to maintain the angle θ with respect to the target.

After illumination, the attacker may need to perform a defensive maneuver that entails an additional angle away from the target, resulting in additional lateral movement. Finally, if it is necessary to run from the adversary to evade a missile shot, a further turn, with associated lateral movement, will be undertaken. These maneuvers are illustrated in Figure D.10.

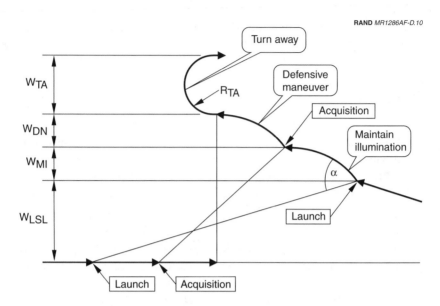

Figure D.10—DCA Post-Launch Lateral Movement

All of these maneuvers, plus an additional initial lateral separation (shown in the figure) imply the need for a width to the combat area. These need to be two-sided, with respect to the target aircraft. Additionally, the target aircraft must itself have an offset from the centerline of the combat area, if only to avoid predictability. These lead to the *notional* formula

$$C_W = 2 \times \{\text{TargetDisplacementFromCenter} \\ + \text{InitialLateralSeparation} \\ + \text{AttackerPostLaunchLateralMovement}\} .$$

In Figure D.10, TargetDisplacementFromCenter is not shown, InitialLateralSeparation corresponds to W_{LSL} (LSL stands for lateral separation at launch), and

$$\text{AttackerPostLaunchLateralMovement} = W_{MI} + W_{DN} + W_{TA} .$$

Table D.7

DCA Lateral Movement Worksheet (Notional Values)

Input/ Output	Variable	Description	Value
Input	RLaunch	Launch range (nm)	25
	LaunchOffset	Lateral offset of attacker at launch (nm)	5
	SpdTgt	kts	575
	SpdAtk	kts	575
	GimbalLim	Gimbal limit of radar, to maintain track before AMRAAM acquisition	60
	Racq	Acquisition range of AMRAAM (nm)	10
	SpdMsl	Average missile speed, ft/sec	2500
	Tdef	Time spent in defensive maneuver, after acquisition	20
	GTurnOff	Gs for turnoff after defensive maneuver	3
Output	y-max	Maximum lateral displacement (from target track) by attacker	12.7

It is important to keep in mind that the maneuvers in the figure are notional and do not define maneuvers that will be executed in all, or even most, DCA engagements. Any individual engagement may require less lateral space. However, the lateral space indicated here must be available to accommodate the widest reasonably anticipated maneuver. Table D.7 shows computations for one notional set of aircraft and missile capabilities.[4]

To compute the lateral requirement for DCA, in addition to inserting real-world values in the worksheet, it is necessary to add several values to y-max (computed as in Table D.7). First, a buffer must be added. Second, the maximum displacement of the target track from the centerline of the area must be added. Finally, the sum of the three factors must be doubled to account for multiple aircraft on both sides of the area centerline.

---

[4]Notional values are used to maintain an unclassified document.

For example, using the y-max value of 12.7 from the example work-sheet, a buffer of 5 nm, and a maximum target track displacement of 10 nm, the total width requirement is computed as

$$2 \times (12.7 + 5 + 10) = 55.4 \text{ nm.}$$

SUMMARY OF DIMENSIONS

Table D.8 provides a summary of the lateral dimension requirements computed using the simulations discussed in this appendix.

Table D.8

Summary of Simulation Results

MDS	Sortie	Length (nm)	Width (nm)	Comments
F-15E	BSA	21.3	10.2	
F-16	BSA	18.0	8.3	
A-10	BSA	14.0	9.2	
F-15E	SAT	25.7	33.0	Does not include DCA orbit; single axis of attack.
F-16	SAT	26.1	30.3	Does not include DCA orbit; single axis of attack.
A-10	SAT	Not simulated	Not simulated	Low-level navigation requires larger area than weapon delivery.
Notional	DCA	Not simulated	55.4	Based on notional example.

DATA LIMITATIONS

This appendix addresses known limitations in the data used for PAF's analysis and embedded in the range and airspace database. Limitations exist in data regarding both requirements and current infrastructure. Each of these is discussed in turn below.

REQUIREMENTS-RELATED DATA PROBLEMS

Sortie Requirements per Pilot

Sortie requirements per pilot, used to calculate required infrastructure capacities, were derived primarily from annual RAP tasking messages. However, RAP messages do not include demands for basic skill sorties, such as AHC. We have largely ignored the demand for such sorties, assuming that these skills are practiced during sorties that are logged in other ways.

A more important problem is that the allocation of commander option sorties is subjective. We had no reliable information on how this is actually done. As a result, we distributed commander option sorties to specified sortie types in proportion to how those types are represented in RAP tasking messages.

A major issue arises because the sortie definitions used for this project do not correspond precisely to RAP sorties. First, we found it necessary to subdivide certain RAP sorties that have different infrastructure requirements, depending on how the sortie is flown. For example, we split SAT sorties into air- and ground-opposed variants, which have significantly different airspace requirements. These variants are just finer subdivisions of RAP sorties, but a mapping is needed that gives the proportion of each RA type. We had to use our best judgment (assisted by ACC) to estimate these proportions.

Additionally, some sorties in our framework "collect" from multiple RAP sorties. These are designated *small multi-MDS engagement* (SMME) sorties. They generally do not appear in RAP but were included in our framework to illustrate both the need for such sorties and to capture the additional infrastructure requirements they would entail. An example of this is AWACS_AA, which trains interactions of air-to-air fighters with AWACS. This sortie includes a fraction of DCA, SAT, and OCA sorties, but has extra infrastructure requirements because of the need for an adjacent AWACS orbit that is properly oriented with the attack axis. Again, we had to use our best judgment to guess what fraction of each sortie type should train with AWACS because this requirement is not in RAP for AWACS or fighter combat crew members.

Number of Pilots

Pilot counts, also used to compute infrastructure capacity requirements, were based on PMAI and crew ratio data rather than actual head counts. This is appropriate because requirements based on the product of a base's PMAI and the MDS' crew ratio should provide a better (and more stable) average estimate of the demand for training infrastructure near a given base than a head count. Unfortunately, the PMAI data we used are possibly outdated and the crew ratio values we used are from multiple sources of varying reliability. Additionally, sortie counts depend on pilot experience levels, which are currently declining. The data for experience levels are based on a recent snapshot of pilot inventory. Similarly, RPI 6 pilots add an additional demand for training; their count is based on a snapshot of the actual inventory. In general, the dates for data for PMAI, crew ratios, experience levels, and RPI 6 are not the same.

Adjustments to Sortie Requirements

In computing the time demand for ranges and airspace, we inflated RAP-derived sortie counts to account for attrition (maintenance and weather cancellations), scheduling inefficiency, and noncontinuation training sorties (see discussion in Chapter Two). We found no empirical data from which to estimate these factors. The factors currently embedded in the range and airspace database should be reviewed and refined, if possible.

Peaks in Demand

Demand is not uniform over a year, but can vary in response to phenomena such as preparation for and recovery from deployment. We assumed level demand throughout the year. However, to maintain appropriate readiness levels, sizing infrastructure supply to service such peaks might be more appropriate than sizing to average demands.

CURRENT INFRASTRUCTURE DATA PROBLEMS

Data describing currently available infrastructure have various problems. Data were collected using a spreadsheet form distributed throughout ACC. This discussion will be limited to data problems that were inherent in the forms (as opposed to problems with the responses), and which as a result limit the analysis that can be performed.

Certain sorties, such as BFM and AHC, require block altitudes, whereas most sorties require a specific altitude range. However, the actual special-use airspace (SUA) floors are specified either as mean sea level (MSL) or above ground level (AGL), with the latter being unsuitable for the evaluation of block altitudes because ceilings are always specified as MSL. In general, both MSL and AGL floors should be provided, or (preferably) either one of these plus an average or maximum SUA ground altitude. Using the average would ignore the effect of widely varying altitudes over the SUA, whereas using the maximum might depict usable altitudes too conservatively.

Opening/closing times and days per week are not as simple as the form would lead one to presuppose. One issue is how to treat cases in which reported opening/closing times are "sunrise/sunset." We replaced sunrise and sunset with 0600 and 1800, respectively, which is reasonable for training requirements spread across an annual cycle. However, seasonal variations in training schedules (caused, for example, by contingency deployments) could make our assumption invalid. Also, when longer hours/days can be prearranged, the instructions ought to indicate the longest workable period. The information of interest is not how much the infrastructure *is* open, but how much it *could be* open to satisfy demand. Even here, workload or funding limitations would presumably limit the maximum average period to something less than the maximum short-term open period.

Yet another problem with opening and closing times is associated with infrastructure (especially military routes) that span time zones. Specifying zulu times in all cases is probably best.

For routes, alternate entry and exit points are not currently usable in the range and airspace database because coordinate information is not supplied.

Composite Ranges and Areas

Ranges and airspace are often designated in sets. Elements of a set may be used individually or may be combined with contiguous elements of the set to produce ranges and SUA with greater lateral or vertical dimensions. These composite ranges and airspace are useful for training in the more space-demanding scenarios. When infrastructure is locally scarce, it is important to avoid double-counting the availability of an individual area that is part of a composite area. Most of the data supplied to us pertain to individual areas, but this is not always the case. For instance, Davis-Monthan AFB reports a single range for the entire Goldwater complex, whereas Hill AFB reports six separate range elements within the Utah Test and Training Range, designated by the restricted airspaces that cover them.

We have defined some composite areas, but our work is based on map data and may not account for unique situations that make it difficult to actually train in a composite area. For instance, some elements may be under different scheduling or air traffic control authorities.

REFERENCES

Chairman of the Joint Chiefs of Staff, *Joint Vision 2010*, Washington, DC: Office of the Joint Chiefs of Staff, undated.

Headquarters United States Air Force, AFI 11-202, Vol. I, *Aircrew Training*, Washington, DC, December 1, 1997.

Kent, Glenn A., and David A. Ochmanek, *Defining the Role of Airpower in Joint Missions*, RAND, MR-927-AF, 1998.

National Imagery and Mapping Agency (NIMA), *Digital Aeronautical Flight Information File*, Washington, DC, August 6, 1999.

Pirnie, Bruce, and Sam Gardiner, *An Objectives-Based Approach to Military Campaign Analysis*, RAND, MR-656-JS, 1996.

Thaler, David, *Strategies to Tasks: A Framework for Linking Means and Ends*, RAND, MR-300-AF, 1993.